WOMEN OF
THE WORLD

EXTRAORDINARY EXPLORERS

EXTRAORDINARY EXPLORERS

WOMEN OF THE WORLD

Women Travelers and Explorers

Rebecca Stefoff

Oxford University Press

New York • Oxford

To my grandmother, Delefern Hubbard—
a traveler and a woman of the world.

Oxford University Press

Oxford   New York   Toronto
Delhi   Bombay   Calcutta   Madras   Karachi
Kuala Lumpur   Singapore   Hong Kong   Tokyo
Nairobi   Dar es Salaam   Cape Town
Melbourne   Auckland   Madrid

and associated companies in

Berlin   Ibadan

Published by Oxford University Press, Inc.,
200 Madison Avenue, New York, New York 10016

Oxford is a registered trademark of Oxford University Press

Design: Charles Kreloff
Picture research: Wendy P. Wills

Library of Congress Cataloging-in-Publication Data

Stefoff, Rebecca
Women of the world: women travelers and explorers /Rebecca Stefoff.
p. cm. — (Extraordinary explorers)
Includes bibliographical references and index.
Summary: Describes some of the explorations and discoveries ,
made by women throughout history, including Ida Pfeiffer,
Isabella Bird Bishop, and Florence Baker.
ISBN 0-19-507687-7
1. Women travelers—Biography. [1. Explorers] I. Title.
II. Series: Stefoff, Rebecca   Extraordinary explorers.
G175.S84 1992
910'.82—dc20   92-7948
CIP
AC

ISBN 0-19-507687-7
1 3 5 7 9 8 6 4 2

On the cover: Marguerite Baker
Harrison shares a meal of wild goat en
route from Istanbul to Persia in 1924.

Printed in the United States of America
on acid-free paper

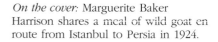

Contents

INTRODUCTION

The Moment of Emancipation

Books about the "Golden Age of Exploration" or the "Great Voyages of Discovery" are books about men. Often the only women mentioned in these histories are Queen Isabella of Spain, who helped pay for Christopher Columbus's trip to the Americas, and Queen Elizabeth I of England, who rewarded Francis Drake with a knighthood when he came home from sailing around the world. For a long time exploration, like many other interesting and adventurous pursuits, was considered man's work.

But there were exceptions. No doubt many women whose stories will never be known have explored new places and made important discoveries. Over thousands of years, the human race has migrated into nearly every part of the world. Each place was once a new place, and in those early days women must sometimes have been the first to cross a mountain range or step onto an island. Perhaps the first person in North America—an Asian nomad who crossed the Bering Strait tens of thousands of years ago—was a woman.

A statue in North Dakota commemorates Sacagawea, a Shoshone woman who crossed western North America with explorers Meriwether Lewis and William Clark in the early years of the nineteenth century. She is one of many women explorers whose exploits are known to us only from legends and tantalizingly incomplete scraps of information.

But in modern times people have used the term "explorer" to mean someone who goes out into the world with a special purpose: to find a place that is unknown to his particular people or culture, to see what is there, and to bring news of it back to his own country. Exploration came to mean the discovery of new lands and seas, the long probing and charting of coastlines and rivers until at last the true outlines of the world's continents could be marked on the map. This work was done by men. But it is not the only kind of exploration.

"The map is not the terrain," wrote Karl von Clausewitz, a German military strategist of the early nineteenth century. He meant that even if you know the location and outline of a place on a map, you still do not know what it is like to stand there. A square of green on a map may be a tropical rain forest where you will sink up to your ankles in mud, be bitten by disease-carrying flies, and marvel at a hundred glorious orchids growing from every tree. Or it may be a dry and level plain of grass, where in the dusty distance you will see browsing herds of elephants and zebras. Filling in the map's outlines, getting to know the terrain—this too is a type of exploration, and women have done a great deal of it.

Women travelers have left countless trails through history. Some of them were religious pilgrims who ventured far from home to see the shrines and holy places of their faiths. One of the first of these was Etheria (or Egeria), a nun from Spain or southern France. In the late fourth century A.D. she not only traveled to Jerusalem and Egypt to visit places mentioned in the Bible but also wrote a brief account of her journey to guide other pilgrims. Her description of the view from atop Mount Sinai is one of the earliest pieces of travel writing by a woman.

Some early women travelers made journeys with foreign explorers. The best-known of these is Sacagawea, a Native American woman of the Shoshone people. She was one of the guides and interpreters for the Lewis and Clark expedition that crossed western North America in 1805-06. Sacagawea joined the expedition in North Dakota, along with her French-Canadian husband, Toussaint Charbonneau. She accompanied Lewis and Clark west to the Pacific Ocean and back again to North Dakota, carrying her infant son, Jean Baptiste, on her back the whole way. Sacagawea's most important contribution to the expedition took place when Lewis and Clark entered the Rocky Mountains and met a group of Shoshone warriors. Because she was able to communicate with the Shoshone, the warriors treated the white explorers as friends and gave them food, horses, and information about the country ahead. Sacagawea appears to have been a brave and level-headed traveler. Unfortunately, we know little about her—just a few details recorded by Lewis and Clark in their expedition notes. Her own thoughts and recollections were not preserved.

Another way that women became travelers was through family relationships. The wives, sisters, and daughters of soldiers and diplomats sometimes accompanied their menfolk to far-off places. Emily Eden, for example, was an Englishwoman who spent six years in India in the 1830s as part of the household of her brother, Lord Auckland, the governor-general of Britain's India colony. Her book *Up the Country*, about an eighteen-month trip into the tiger-infested foothills of the Himalaya Mountains, was published in 1866 and became a classic of travel writing. Yet Eden thoroughly disliked traveling. She came to India only out of a sense of duty to her brother, and she eagerly counted the days until she could return to England.

A fourth type of woman traveler was the emigrant who left her homeland to live in another country. Emigrant women may not have been explorers in the strict sense of the word, but they often *felt* like explorers when they left the known and familiar world behind to begin a new life in a strange land. One emigrant woman who kept a record of her experiences was Susanna Moodie. She and her husband left England for Canada in 1832 to settle in the backwoods of Ontario. *Roughing It in the Bush: Or, Life in Canada*, published in 1852, tells the story of their first years as homesteaders, during which they suffered from bitter weather in winter, swarms of mosquitoes in summer, and lean times while they learned to live off the land. Eventually, though, the Moodies adapted to their new home. When she died at age eighty-one Susanna Moodie was one of Canada's most respected novelists, but her best-known book was her tale of emigrant life in the back country.

Thousands of women traveled in the roles of pilgrims, guides, family members, and emigrants. But in the nineteenth century a new class of woman traveler appeared in Europe and the United States. These women left home not to be with their men or out of necessity but simply to please themselves, and they ventured beyond the destinations that were considered safe and respectable for travelers. They went to places where no other woman from the Western world had gone before—and sometimes to places where no man had gone. The more distant and dangerous the destination, the more eager these women were to make the trip. They claimed the right to see the world on their own terms, and they also laid claim to the title "explorer."

Women travelers and explorers emerged at a time when women in Europe and America were reaching toward emancipation, or freedom, in many aspects of life. Women were regarded by the law, by their churches, and by society in general as not having the same rights and responsibilities as men. Their opportunities to study and work were limited; the household was their domain, not the larger world of ideas and events. Women were expected to be good wives and mothers. If they did not marry, they were expected to keep house for their male relatives or quietly earn a living in some genteel way, perhaps by giving drawing lessons or doing embroidery.

Not all women accepted this view of things. Many of them began to challenge the rules. They demanded the right to vote, they sought recognition in the arts and sciences, and they looked for a new sense of liberty in their lives. Some of them found that liberty in travel and exploration.

The women travelers of the nineteenth century were exploring more than just geography. They were also exploring women's place in the world. By showing that a woman could survive the snows of Tibet or the jungles of Borneo as well as any man, they enlarged the world's awareness of what all women can do. They helped prepare the way for the later achievements of women in every field, not just in exploration.

The nine women whose exploits are recounted in this book span more than a century and a half of travel and exploration. They are from five different countries and a variety of backgrounds. Some of them were married; some remained single. Most of them traveled alone, often in perilous places. Each of them was a true explorer, going someplace where no one had gone before or doing something that no one had done before. The details of each woman's adventures are well documented, and all of them left accounts of their travels that have encouraged other explorers, both women and men. Finally, all of these women shared one vital thing: the love of travel itself, not just as a passage to a destination but as an experience with its own meaning and magic.

The world beckoned to these women, and they responded, even though many of them paid a high price for their travels. As the twentieth-century traveler Freya Stark wrote, "The beckoning counts, and not the clicking latch behind you: and all through life the actual moment of emancipation holds that delight, of the whole world coming to meet you like a wave."

CHAPTER 1

Ida Pfeiffer: Twice Around the World

Ida Pfeiffer was born Ida Reyer in Vienna, Austria, in 1797. As a child she daydreamed about visiting China and India and other faraway lands. A few European women *did* travel at that time. The wives of some missionaries or government officials accompanied their husbands to foreign posts, and a handful of aristocratic, wealthy women toured the out-of-the-way parts of Europe and the Middle East. The idea of an ordinary housewife of limited means traveling alone all the way to the other side of the world, for no better reason than because she wanted to see what it was like— such a thing was unheard of!

But Ida Pfeiffer did not let that stop her. Without any fuss or bother, she simply went where she wanted to go. She was the first woman to make traveling her life's work, and the books she wrote about her journeys inspired other women to explore the world. Pfeiffer paved the way for a later generation of women travelers, although few of them covered as much ground as she did.

Ida Reyer grew up in an eccentric household. She had five or six brothers. Her father felt that she should have been a boy, too, so he had her dressed in boy's clothes, and he gave her the same training and education that he designed for his sons. This training was supposed to make them strong and independent as well as to teach them academic subjects. For example, the children took long walks in the countryside and played outdoor games in all kinds of

Although she seems in this portrait to be a demure middle-class housewife, Ida Pfeiffer had the soul of a reckless adventurer. In the course of her travels she rode a camel across the deserts of Persia and hobnobbed with headhunters in Borneo.

weather, and Ida shared these strenuous activities with her brothers. Such behavior was generally forbidden to girls at that time, but Ida was treated just like a boy. Confusing as it was, this early upbringing probably helped form her courage and strong will—a strength that sometimes bordered on foolhardy stubbornness.

Ida Reyer's father died when she was nine, but it took years for her mother to persuade the girl to give up her trousers and start wearing petticoats and dresses. Perhaps young Ida realized that the women's clothing of her time, with its stiff corsets and long, heavy skirts, would limit her freedom of movement. But at the age of thirteen she reluctantly began dressing like a girl. Piano lessons were expected of a well-brought-up young lady; she agreed to take them, but she hated them.

She received her education at home from a private tutor, and when she was seventeen she fell deeply in love with this young man. He loved her, and they wanted to marry, but Reyer's mother would not permit it—she wanted a better match for her daughter than a poor and humble tutor. Reyer spent the next few years in a quiet, unhappy frame of mind, steadily pressured by her mother to make an advantageous marriage. Finally she agreed to marry a Dr. Pfeiffer, a widower who was much older than she. He was a lawyer in the city of Lemberg and held an important position in the Austrian government. They were married in 1820; Reyer was twenty-two.

The Pfeiffers' marriage was not particularly unhappy, and they had two sons together. But soon after they were married Dr. Pfeiffer made enemies in high places. He learned that several government officials had committed misdeeds, and he revealed these offenses to the authorities. The guilty men were thrown out of office, but their vengeful friends created trouble for Pfeiffer. He lost his government appointment and he was never again able to find work.

The Pfeiffer family, which had enjoyed an expensive way of life, now found itself poor. Ida Pfeiffer tried giving music and drawing lessons to earn money, but finally she had to ask her brothers to help pay for her sons' education. Her mother died in 1831, leaving her a small inheritance—just enough so that she could support herself and educate the two boys. In 1835 she separated from her husband, who was elderly and wanted to live in Lemberg with his son from his first marriage. Ida Pfeiffer and her sons moved back to Vienna.

By 1842, both of the boys had established their own homes. Pfeiffer was forty-five years old and free of family responsibilities. She decided that it was time to make her childhood dreams of travel come true.

Her friends were shocked when she told them that she planned to travel alone. She replied that she was mature and self-reliant enough to take care of herself. She had very little money, but, as she later recalled, "I was determined to practice the most rigid economy. Privation and discomfort had no terrors for me." She announced that she would visit the Holy Land (present-day Israel, especially the region around Jerusalem). Because such a journey was a sort of religious pilgrimage, it was considered somewhat respectable. Pfeiffer did not tell her friends, however, that after visiting the Holy Land she intended to wander around in Egypt, which was more dangerous and unfamiliar than the Holy Land. Knowing that death from disease or from a murderous attack was quite likely, she wrote out her last will and testament. Then she set out, not expecting to return.

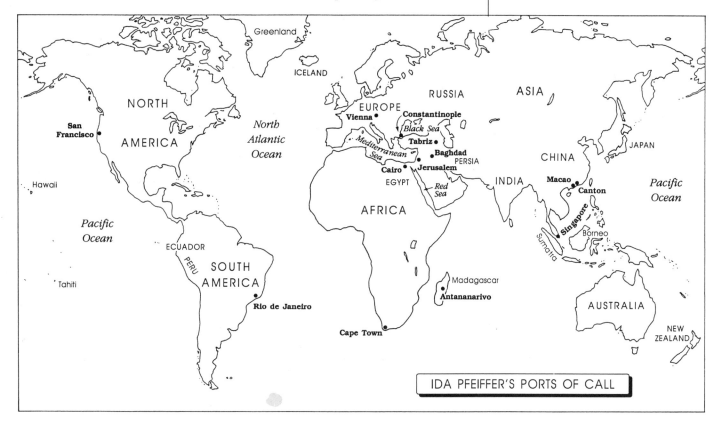

IDA PFEIFFER'S PORTS OF CALL

She was back in high spirits nine months later. She had sailed down the Danube River through southeastern Europe to the Black Sea. Then she crossed the Black Sea to the metropolis of Constantinople, in Turkey, the gateway to Asia. From Constantinople she went to Jerusalem, and from there to Cairo, Egypt. After spirited bargaining with a camel-driver—whom she was sure had cheated her—she learned to ride one of those temperamental one-humped dromedaries. She strolled in blazing heat around the Pyramids of Giza and the nearby Sphinx, and she crossed the desert to the Isthmus of Suez, where seventeen years later the French would begin building a canal to connect the Mediterranean with the Red Sea. At last she sailed across the Mediterranean and came home by way of Italy. She could hardly wait to start another trip.

Pfeiffer crossed the Egyptian desert to see the Isthmus of Suez. Today this narrow land bridge between Africa and western Asia is cut through by the Suez Canal, but in Pfeiffer's time no ship could pass from the Mediterranean Sea to the Red Sea.

Throughout her journey Pfeiffer kept a diary, which she read to her friends upon her return. They suggested that she turn it into a book, and it was published in 1846 as *Visit to the Holy Land, Egypt, and Italy*. Even before the book appeared she had used the money from her publisher to pay for a second trip, this time to Iceland—an adventurous choice.

Iceland is considered to be part of Europe, but it lies far out in the North Atlantic Ocean. In 1845 a visit to Iceland was an arduous undertaking. Pfeiffer reached the island on a cramped, damp, cold ship, and once there she had to bargain for transportation in pony carts. Some European travelers—mostly British—had toured Iceland before her, but they were rich enough enough to equip large expeditions and to travel in some degree of comfort, with servants

and a few luxuries. Pfeiffer, as usual, traveled alone and on a tight budget.

Earlier travelers had painted a romantic picture of Iceland as a noble land of spiritual people who lived close to nature. After six months in Iceland, living like the Icelanders, Pfeiffer took a harsher view: She found the people crude, the homes dirty, and the culture primitive. Among other things, she complained about the fleas she found in almost every bed and about the boring meals she was served—mostly porridge and fish. She published her observations as *Journey to Iceland, and Travels in Sweden and Norway* (1846). With the money this book earned her, plus what she raised by selling to museums the rocks and plants she had collected in Iceland, she was able to scrape together funds for a more ambitious venture. This time she planned to go around the world.

Pfeiffer departed in 1846 on a Danish vessel bound for South America. She landed in the Brazilian port of Rio de Janeiro, whose scenic loveliness had been praised by previous visitors. But she was repelled by the filth and poverty she saw in her first glimpse of the tropics, and she wrote, "There is nothing to offer compensation for the disagreeable and repulsive sights that meet your eyes at every turn." Soon after her arrival she was attacked by a knife-wielding

Pfeiffer's first book, based on her travel diary, helped pay for her second trip. From then on she supported herself by writing about her journeys.

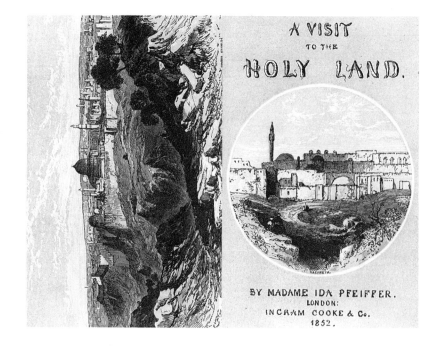

brigand, but fortunately she was not seriously hurt. After that, she carried a pair of pistols.

She was unimpressed with the "civilized" parts of South America, the cities and plantations, so she hired a guide and went into the rain forest to investigate the Indians. Her report on this excursion reveals two sides of Pfeiffer the traveler. One side is delighted by new sights and experiences. The other is critical and intolerant of other peoples and cultures.

Pfeiffer was enchanted by the natural beauty of the forest. Gazing at the towering trees festooned with orchids and twining vines, and at the brilliant plumage of the multicolored tropical birds that flitted to and fro above her, she felt—like many other travelers—that she had entered a new and different world, a world of dream or imagination. "It seemed to me that I was taking a ride in fairy-land," she wrote. "I was over happy, and felt every exertion I had made most richly rewarded."

But the enchantment ended when she reached her goal, a tribe of Puri Indians. Now the critical and intolerant Pfeiffer emerged. She thought that the Indians' houses—huts roofed with palm leaves and open to the air on three sides—were ridiculously primitive, even though they were well suited to the climate. She thought the Indians were ugly and stupid (though she did admire their hunting skills). They greeted her with a dance, but she hated the loud music and the yelling. In short, she had no doubt at all that as a European and a Christian she was superior in every way to the Indians, whom she called "savages."

From South America Pfeiffer set sail across the Pacific Ocean for China. Along the way she spent several weeks in Tahiti, a place very few Western women had visited; she was scandalized by the open, carefree sexual behavior of the Tahitian women.

Pfeiffer was very brave—some might say very foolish—to travel in China in the late 1840s. Great Britain and China had been at war from 1839 to 1842, and Europeans of all sorts were highly unpopular in China. Furthermore, China's peasant population was seething with revolt against the ruling family; in 1850, just a few years after Pfeiffer's visit, the unrest erupted in the Taiping Rebellion, a bloody civil war that raged across China for more than a decade and left twenty million people dead.

She entered China by way of Macao, a Portuguese colony on the south coast. She first saw Chinese people in Macao harbor, and her comment once again shows her two sides—first the pleasure she took in seeing the world for herself, and then her tendency to see other people in terms of ungenerous stereotypes. "A year ago," she wrote, "I should have little thought there was any chance of my becoming acquainted with this remarkable country, not merely from books but in my proper person; that the shaven head and long tails and cunning little eyes, as we see them in pictures and on tea-chests, would have presented themselves in living form before me."

Pfeiffer traveled on a junk, a traditional Chinese cargo boat, to the city of Canton (today called Guangzhou). She learned that Louis Agassiz, a well-known Swiss-American biologist, was in the city on a scientific mission, and she decided to call on him. She later wrote that as she walked through the crowded streets, "old and young had called and hooted after me, and pointed their fingers, and run out of the shops." By the time she reached Agassiz's quarters, she was followed by an aggressive mob. Agassiz, she said, was "excessively surprised" to see a solitary white woman come walking up, and he told her that it was dangerous for her to be in Canton. But Pfeiffer was too stubborn to be driven away by fear. She felt that "there was nothing to be done but to put a good face on the matter, and I therefore marched fearlessly on; no harm happened to me."

Although she was occasionally insulted or frightened by the Chinese, Pfeiffer boldly made many excursions. She was most eager to leave the part of Canton where the Europeans lived; she wanted to walk through the Chinese quarter and along the city walls, "an attempt no woman had yet ever ventured to make." She did it, although she dressed in men's clothes for safety. And she took every opportunity to observe Chinese life and customs, taking copious notes on everything from household furniture to burial customs. She formed a low opinion of the people: "a baser, falser, crueller people than the Chinese I never met with," she wrote.

Her next stop was India, where she had a better time. She spent some months roaming around quite alone and happy, now and then joining in such pastimes as a tiger hunt or a Hindu festival. She traveled with almost no baggage. Unlike nearly every European who went abroad, she carried no medicines of any kind. Often she did not even have a blanket. Her only food supplies were "a leathern [sic] bottle for water, a small pan for cooking, a handful of salt, and some bread and rice."

Throughout her travels, Pfeiffer expected the people she met to feed her and offer her shelter, and usually they did. She ate roast monkey and corn with the South American Indians, boiled rice and eggs with the Hindu villagers. She grew offended and sarcastic on the rare occasions when people did not offer her hospitality— although she does not say what she would have done if a turbaned Hindu had appeared on her doorstep in Vienna and demanded a meal!

Leaving India, Pfeiffer traveled north through Mesopotamia (today called Iraq) to the ancient city of Baghdad. There she joined a camel caravan for a 300-mile (480-kilometer) journey across the desert to the city of Mosul. She was eager to visit Persia (present-day Iran), so she followed the caravan to Tabriz in northern Persia— first sending all her notes home to her sons in case she was killed by bandits or hostile tribespeople. But she arrived in Tabriz without trouble. The British consul there was amazed to see her, having no idea that a woman could travel alone in the region without even knowing the native languages.

For the time being, however, Pfeiffer had had enough of traveling in countries that she considered backward and benighted. She

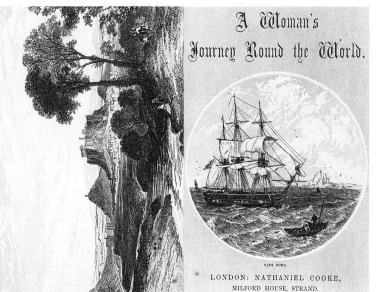

The book she wrote about her first round-the-world trip made Pfeiffer a celebrity—and brought her many invitations and offers of assistance for her second trip.

attached herself to another caravan that was headed north toward Russia, where the people were Christians, and she told herself that it would be good to be among "civilized" folk again. To her dismay, she found conditions in southern Russia to be just as "barbaric" as those in Persia. As she walked along with the caravan, armed Russian soldiers grabbed her and carried her off. They had taken her for a spy. She was held overnight while they checked her identity, and then she was unceremoniously released. Pfeiffer was highly indignant. "Oh you good Arabs, Turks, Persians, Hindoos! How safely did I pass through your heathen and infidel countries; and here, in Christian Russia, how much have I had to suffer in this short space," she wrote.

The rest of her trip through Russia was no better. She grumbled at length about the delays, rudeness, and red tape she encountered before she finally managed to leave Russia and make her way westward through Turkey, Greece, and Italy. Pfeiffer reached home in November 1848. She had been away for nineteen months and had traveled 35,000 miles (56,000 kilometers) by sea and 2,800 miles (4,480 kilometers) by land. She wrote up her adventures in *A Lady's Voyage Round the World,* the first of her books to be translated into English.

This book made her famous. When she announced that she intended to travel around the world again beginning in 1851, she received invitations to visit from Europeans stationed in outposts all over the world. Railroad and steamship companies competed with each other to offer her free tickets.

Pfeiffer sailed from London to Cape Town, a British colony at the southern tip of Africa. She had considered venturing into the interior of Africa, but decided that she could not afford it, so she went east to Singapore, intending to take ship to Australia. But in Singapore she changed her plans again—and embarked on the true exploring phase of her career.

She went to Sarawak, a territory on the northern coast of the huge jungle island of Borneo. Borneo was one of the least-known places in the world, but since 1842 Sarawak had been governed by James Brooke, a British adventurer, and a few Europeans had begun to visit the region.

Pfeiffer spent six months in Borneo, eagerly taking in all the sights and sounds of the mountainous, almost impenetrable tropical rain forest. Among the inhabitants of Borneo were the Dyaks, who practiced ritual headhunting and were thought by Europeans to be especially "savage" and fierce. They lived in the forest in long thatched houses that sheltered 100 or more people. Shrugging off the warnings of Brooke and other advisers, Pfeiffer insisted on traveling to the communal longhouses of the Dyaks—including the "free Dyaks," tribes in the interior who had not been brought under British influence. To reach them, she donned what she called "a simple and appropriate costume" consisting of trousers (with a petticoat over them), a jacket, and a huge bamboo hat. She traveled by boat on the crocodile-infested rivers when she could, but she was forced to cross several steep mountains on foot through dense jungle.

Surprisingly—in view of her generally low opinion of native peoples—Pfeiffer

"I saw all I possibly could, never missed an opportunity of mingling with the people, and carefully noted down all I saw," wrote Pfeiffer of her stay in China. This illustration of a Chinese festival called the Feast of Lanterns appeared in *A Lady's Voyage Round the World*.

In India, Pfeiffer visited the Taj Mahal, a marble palace built in the seventeenth century as a memorial to a dead queen. Pfeiffer called it "the most magnificent treasure of all India," and it is still one of the country's leading tourist attractions.

liked and admired the Dyaks. Although she admitted that she "could not look without horror" at the row of heads that hung in one of the longhouses she visited, she was uncharacteristically fair-minded. "I shuddered," she wrote, "but I could not help asking myself whether, after all, we Europeans are not really just as bad or worse than these despised savages? Is not every page of our history filled with horrid deeds of treachery and murder?" She decided that the atrocities of the Dyaks were no worse than the religious wars of Germany and France, the European conquest of the Americas, or the Spanish Inquisition.

The Dyaks treated Pfeiffer courteously, and she later wrote, "I should like to have passed a longer time among the free Dyaks, as I found them, without exception, honest, good-natured, and modest in their behavior. I should be inclined to place them, in these respects, above any of the races I have ever known." Borneo seems to have set a new standard for Pfeiffer; from then on, all the peoples that she encountered were compared unfavorably not just with Europeans but also with the noble Dyaks.

After leaving Borneo, she went to the Dutch East Indies, the cluster of islands in southeastern Asia that now make up the nation of Indonesia. On the island of Sumatra she decided to look in on the Batak, a people who were known to be cannibals and had never allowed a European to travel into their territory. Once again her European hosts made strenuous efforts to discourage her, once again she insisted—and once again she returned safely from a dangerous trip.

The Batak were undoubtedly astonished by Pfeiffer's visit. They treated her as a great curiosity, passing her from tribe to tribe, and occasionally they threatened her. She saw a ritual dance that mimicked the killing and eating of a mock human victim. "Play as it was, though, I could not witness it without some shuddering, especially when I considered that I was entirely within the power of these wild cannibals," she later recalled.

At one point Pfeiffer found herself in the midst of a crowd of Batak who showed by gestures that they wanted to kill and eat her. She was frightened, but she had prepared a joke in the Malay and Batak languages for such an occasion. Desperately hoping that the Batak would understand her, she remarked that she was too old and tough to make good eating. As she had hoped, this amused them, and the moment passed. But the danger remained real.

Although Pfeiffer had hoped to visit a distant Batak tribe that was said to be ruled by a woman, she decided that it would be more prudent to turn back while she could. The Batak did not let her leave at once, but eventually she escaped unharmed. She was the first person to report on their customs and way of life.

Tired and rather ill after her Indonesian hardships, Pfeiffer sailed from the East Indies to San Francisco. She toured the mining camps of the California gold rush and stayed for a time with a Native American people called the Yuba. Then she visited the Pacific coast of South America—the Andes Mountains of Ecuador and Peru—before crossing the United States. She returned home in 1855 after a four-year absence.

By this time Pfeiffer was a figure of international renown. Her book *A Lady's Second Journey Round the World* was a best-seller. She was befriended by Alexander von Humboldt, an influential and respected German scientist who had been a noted traveler in his youth, and with his support she was elected to the geographical societies of Berlin and Paris. But the Royal Geographical Society of Great Britain, the most prestigious such organization in the world, refused to admit her because she was a woman.

Pfeiffer looked around her for the next challenge. She decided to visit Madagascar, a large island off the southeast coast of Africa. Many unique species of plants and animals live in Madagascar, and for this reason geographers consider it one of the most exotic and interesting places in the world. In Pfeiffer's time it was also one of the most dangerous. In 1845 Queen Ranavalona, the bloodthirsty tyrant who ruled the island, had killed or driven away almost all the Europeans there. Alexander von Humboldt urged his friend not to go to Madagascar, but as usual she insisted on having her own way. This time, however, her luck ran out.

When Pfeiffer arrived in Madagascar, she found herself a virtual prisoner of the queen in the capital city of Antananarivo. She had to play the piano to amuse the queen; luckily, she remembered some of the music she had been made to learn as a teenager. There were only six other Europeans in Madagascar, all in Antananarivo. They had dreamed up a somewhat confused plot to overthrow Queen Ranavalona and turn the island over to one of her sons, and Pfeiffer unwisely became involved in this conspiracy. The queen discovered the plot and promptly threw the Europeans into prison.

The conspirators expected to be killed. They witnessed orgies of execution in which the suspicious and violent queen poisoned hundreds of her subjects or threw them to crocodiles. But finally the Europeans were simply ordered to leave the island. A guard of soldiers conveyed them to the coast. This nightmarish journey, during which the unlucky conspirators were frequently imprisoned and ill-treated, lasted for fifty days. Pfeiffer became gravely ill with a tropical disease. Finally the prisoners escaped by ship to the nearby British island colony of Mauritius. When Pfeiffer was well enough to travel, she was sent home to Vienna, but she did not regain her health. She died in October 1858 of what her obituary called "Madagascar fever."

In some ways it is hard to admire Ida Pfeiffer. Lacking the ability to judge others on their own terms, she held rigid and generally harsh views about other peoples and cultures. At times it seemed that she traveled only to criticize. On top of that, there was something self-destructive in her headlong refusal to listen to well-meant advice. She seemed sometimes to be *too* determined for her own good.

But perhaps Pfeiffer needed all her stubbornness to make her way alone in places that many people thought were off limits to women. She overcame the problem of poverty, and she pioneered independent travel for women; she also showed that it was possible for a woman to make a living as a professional traveler and travel writer. If she was sometimes arrogant, she was also immensely energetic and curious. Above all, she had what she called "an insatiable desire for travel"—and the strength of will to satisfy that desire.

CHAPTER 2

Isabella Bird Bishop: Escape from Civilization

The second half of the nineteenth century saw a rapid expansion of British influence around the world. Great Britain claimed colonies all over Asia and Africa, and British missionaries, merchants, and government officials carried their various sorts of business to all corners of the empire.

Travelers in growing numbers also spanned the globe in this imperial era. Although for the most part women were still expected to tend the hearth at home, the idea of a woman traveling no longer seemed quite as outlandish as it had during Ida Pfeiffer's youth. A number of intrepid, adventurous women roamed the world during the Victorian period (the reign of Queen Victoria of England, from 1837 to 1901). But no traveler was bolder, more free-spirited, or more renowned than Isabella Bird Bishop.

Isabella Bird (Bishop was her married name) was born in 1831 in Yorkshire, England. Her father was a Church of England clergyman. Under his influence she became a devout member of the church and a Sunday School teacher. She was so deeply attached to her faith that she was intolerant of all other religions, believing them to be harmful superstitions.

Isabella Bird Bishop wears one of her travel souvenirs— ceremonial robes in the style of the Manchus, the ruling dynasty of China. In many ways Bishop was more at home in far-off parts of the world than in her native Britain.

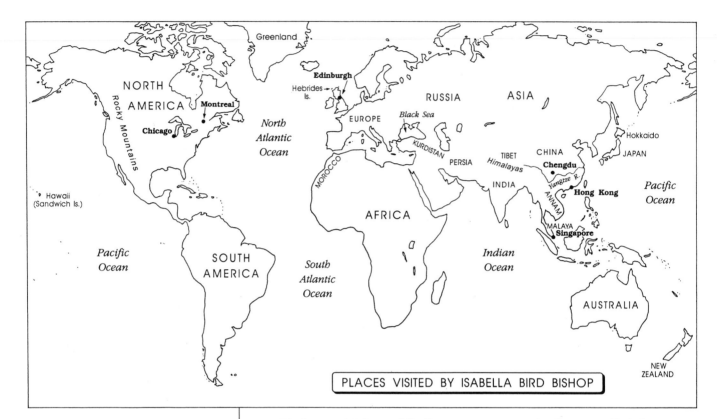

PLACES VISITED BY ISABELLA BIRD BISHOP

Bird was sickly and frail as a youngster. At the age of eighteen, she had a tumor removed from her spine, but her health and spirits did not improve. She could hardly walk or even get out of bed. Several years later the family doctor recommended travel to perk her up—in that era sea air and a change of scene were often prescribed for ailments that did not respond to medicine. So Bird's father gave her a modest sum of money and told her she could travel for as long as it lasted. In 1854 she departed by steamship to visit a cousin in Canada.

As soon as her trip began, Bird experienced a remarkable recovery. Her doctor may have pictured her lying sedately in a deck chair while she steamed across the Atlantic and back again, but instead she found herself filled with energy. For the first time in her life she was on her own—a liberating experience for a twenty-three-year-old Victorian woman—and she found that she was able to take care of herself very well. Bidding her cousin farewell, she set out for a three-month tour that took her to Montreal, Toronto, Chicago, and New England. She traveled by train and along bumpy, rutted roads by stagecoach, and she had a wonderful time. Her family enjoyed her lively, colorful letters. When she came home they suggested that she turn them into a book. She did so, and *The Englishwoman in America* was published in 1856.

Bird's trip had greatly improved her health, but at home she once again grew weak, ill, and depressed. At her father's suggestion, she made several more trips to the United States and Canada to study the state of religion in those countries, but her health did not improve. She suffered from periods of deep depression and was unable to sleep. Nevertheless, although she was bedridden part of the time with back pain, she wrote a number of magazine articles on religious subjects and managed to travel through the Hebrides, a group of Scottish islands. After her parents' deaths, she settled with her beloved sister Henrietta in a cottage on one of these islands.

By 1872 Bird grew so ill and gloomy that a doctor recommended another sea voyage. The patient obediently set off for Australia and New Zealand. She was forty years old, and the adventure of her life was about to begin.

She did not think much of either Australia or New Zealand. She found the landscapes dusty and the people too fond of alcoholic beverages. She booked passage for San Francisco on a leaky, condemned old hulk called the *Nevada*. In the middle of the South Pacific Ocean the ship was caught in a hurricane that raged for twelve hours. Bird was electrified with excitement and compared the experience to falling in love. "At last I am in love," she wrote to her sister. "And the old sea-god has so stolen my soul that hereafter, though I must be elsewhere in body, I shall be with him in spirit."

All of Bird's boredom, illness, and exhaustion seemed to vanish in the combined danger and beauty of the storm. She felt newly awakened to life, and she realized then that travel was a form of freedom. She explained her passion for travel to Henrietta this way: "To me it is like living in a new world, so free, so fresh, so vital, so careless, so unfettered, so full of interest that one grudges being asleep." She devoted the rest of her life to exploring the "new world" of travel. Whenever she returned home she soon became restless and before long was packing for her next trip.

Bird got off the rusty *Nevada* in Hawaii (sometimes called the Sandwich Islands in the nineteenth century). She spent a glorious six months traveling from island to island, touring the plantations and forests on horseback, climbing volcanoes, and peering over cliffs and waterfalls. Women at that time were expected to ride sidesaddle, with both legs on the same side of the horse so that they would remain modestly covered by their long skirts. But Bird saw Hawaiian women riding astride their horses like men did, so she decided to do the same. She had a set of riding breeches made in a bold plaid, although she wore a skirt over them to keep from looking entirely scandalous. As soon as she switched from a "lady's" saddle to a man's saddle, she found that for the first time she could ride without pain in her back.

What Bird liked best was traveling light and on her own, as she described it several years later in her book *Six Months in the Sandwich Islands:*

> This is the height of enjoyment in travelling. I have just camped under a *lauhala* tree, with my saddle inverted for a pillow, my horse tied by a long lariat to a guava bush, my gear, saddle-bags, and rations for two days lying about, and my saddle blanket drying in the sun....The novelty is that I am alone, my conveyance my own horse; no luggage to look after, for it is all in my saddle-bags; no guide to bother, hurry, or hinder me.

From Hawaii, Bird moved on to the United States, where she went alone, on horseback, into the Rocky Mountains. She roamed through California and Colorado, visiting mining camps and ranches, learning to drive wagons and herd cattle. She climbed peaks, rode through blizzards, and was snowed in for several months in a mountain cabin.

She also fell in love—with a hunter and scout named Mountain Jim Nugent, who guided her around Colorado. In his company she scrambled up some 15,000-foot (4,575-meter) Rocky Mountain summits. The vigorous outdoor life and the sublime scenery, as well as Nugent's companionship, exhilarated Bird. She described to Henrietta a day when she and Nugent made a difficult ride up into the mountains: They stood looking out over the world below "in the splendour of a sinking sun, all colour deepening, all peaks glorifying, and shadows purpling, all peril past." For Bird, these moments of supreme beauty and freedom were the essence of travel. They were rare and hard to achieve, but they were the happiest moments of her life.

Nugent returned Bird's affection and tried to persuade her to stay in Colorado. But although she was strongly attracted to him, she decided to leave, partly because she recognized the flaws in his hot-tempered and rough character and partly because she feared that marriage would limit her newfound freedom. The decision was probably a wise one. Six months after she left, Nugent was shot to death in a quarrel on one of the ranches where Bird had stayed.

Bird returned home in 1874 after a trip that had lasted several years and carried her around the world. *A Lady's Life in the Rocky Mountains,* her book about her Colorado adventures, was published in 1879 and was enormously popular.

Back home in Scotland, Bird took a class in botany. One of her classmates was Dr. John Bishop, a gentle, soft-spoken physician ten years younger than she. The two became friends, and soon Bishop was the Bird sisters' doctor. He fell in love with Isabella and made the first of many marriage proposals in 1877. Bird turned him down and promptly set off on a trip to Japan. She hired a Japanese guide and spent seven months traveling up and down the mountainous island kingdom, energetically searching out the most remote villages and temples.

In the American West as in Hawaii, Bird scandalized prim and proper citizens by wearing a divided skirt over long bloomers and sitting astride her horse like a man.

Convinced of the superiority of the British brand of Christianity, Bird was scornful of the Japanese religion. She also had little use for the city-dwelling aristocrats, whom she regarded as arrogant and decadent. But she liked the humble people of the Japanese countryside—mostly because they did not interfere with her activities. She reported with pleasure that "a lady with no other attendant than a native servant can travel, as I have done, for 1,200 miles [1,920 kilometers] through little-visited regions, and not meet with a single instance of incivility or extortion."

Bird did some genuine exploring during her visit to Japan. She went to Hokkaido, the northernmost island. Hokkaido is the home of the Ainu people, Japan's original inhabitants, whose language and religion were unlike those of the other Japanese. Very few Europeans had visited the Ainu before, and probably no European woman had done so. Bird stayed in villagers' huts and climbed through dense forest to observe the shrines of their ancient religion, which was based on the worship of the bear. Two Ainu took her on a canoe ride up the Sarufutogawa River, and she wrote later, "I had much the feeling of the ancient mariner—

> We were the first
> Who ever burst
> Into that silent sea.

For certainly no European had previously floated on the dark and forest-shrouded waters."

The lines that Bird quoted from Samuel Taylor Coleridge's 1798 poem "The Rime of the Ancient Mariner" show that she felt herself to be entering an unexplored land. Now she learned to enjoy the sensation not just of traveling but of going where few, or none, had gone before her. She prided herself on never taking the easy route; she always looked for some out-of-the-ordinary challenge. When she published her two-volume account of the trip, she gave it the title *Unbeaten Tracks in Japan: An Account of Travels on Horseback in the Interior.*

After leaving Japan she stopped briefly in Hong Kong, South China, Annam (now called Vietnam), and Singapore. She was offered an opportunity to go to Malaya on a Chinese steamer, and without hesitation she accepted. Malaya, consisting of the Malay Peninsula and a number of large nearby islands, had come under British influence, but it was almost entirely unexplored. Although the coastlines were fairly well known, there were no maps of the interiors of the islands. Bird dauntlessly made her way through several Malayan states, relying on Malay guides and, when necessary, riding elephants across muddy rivers. "This mode of riding," she reported to Henrietta, "is not comfortable."

On one occasion her elephant waded into water so deep that the animal was almost completely submerged, "not a bit of his huge bulk visible" except the end of his nose, poking up from the water ahead. She herself was immersed up to the neck, but the water was "nearly as warm as the air" and she enjoyed the experience of riding "some distance up the clear, shining river with the tropic sun blazing down upon it."

She had other odd experiences in Malaya as well. She stopped at one government station operated by a British administrator, but he was away on business. In his absence his Malay servants set a formal table for Bird's dinner. She was surprised to see three chairs at the table, as she was the only person at the station other than the servants, but she was even more surprised when two apes, a large one and a small one, were brought in to serve as dinner companions. All three of them dined on fruit and eggs, and Bird remarked that the apes had very good manners, except that they occasionally grabbed food without waiting for it to be passed. She reported this and other adventures in letters to Henrietta, which she later made into a book called *The Golden Chersonese and the Way Thither* (*chersonesus* is Latin for "peninsula," and people with classical educations sometimes called the Malay Peninsula the Golden Chersonese).

Not long after Bird returned home, Henrietta died, and Bird suddenly felt very lonely. The patient and affectionate Dr. Bishop renewed his offer of marriage, and this time Bird accepted. They were married in 1881. By this time she had become particularly

A storehouse used by the Ainu people in Hokkaido. Bird lived among the Ainu, Japan's earliest inhabitants, and gave a detailed account of their culture and customs, which were almost entirely unknown to Europeans at the time.

interested in Tibet and the other ancient and little-known countries between India, Russia, and China; she dreamed of traveling there. Dr. Bishop once said, "I have only one formidable rival in Isabella's heart, and that is the high tableland of Central Asia." The Bishops never traveled together, however. Some biographers feel that Isabella did not really want to travel if she could not have the pleasure of solitude. In addition, she became ill again and had to undergo a second operation on her spine.

John Bishop had waited five years to marry Isabella Bird, but their marriage, though it was loving and devoted, did not last even that long. He died in 1886. Isabella Bird Bishop spent a year or so trying to act the part of a sedate, respectable widow in her fifties—she gave French lessons and drawing lessons and tried to take up nursing—but by 1888 she was packing her bags for another trip to Asia.

She went to Pakistan and India, as always traveling light. Although she planned to be on the road for an indefinite length of time, all

Armed with a parasol, Bird takes her first elephant ride in Malaya.

her supplies were contained in four very small boxes and a waterproof bag. The only food she carried was a little tea, dried soup, and saccharin—she bought everything else as she went.

Bishop found India annoyingly "civilized" under British rule. She complained that "there was no mountain, valley, or plateau, however remote, free from the clatter of English voices." So she decided to visit a remote kingdom called Ladakh in the Himalaya Mountains. Ladakh had come under the control of the Indian province of Kashmir, but it was inhabited by Tibetan people and was often called called Little Tibet. It could be reached only by a daring ride across treacherous, snow-filled passes and through flood-swollen streams.

She set off in high spirits with four native companions: an interpreter, a groom for the horses, a porter to carry provisions, and a soldier who was assigned by an Indian maharajah to protect her. After Bishop had been traveling with him for several months, the soldier was arrested on a murder charge, and she calmly noted that "an attendant of this kind is a mistake."

Bishop spent some time in Ladakh and made several side trips to the camps of nomadic Tibetan tribes. One of her journeys required her to ride a yak—a shaggy buffalolike beast "with an uncertain temper"—over a 17,930-foot (5,468-meter) pass. She had a grand time and felt more alive and energetic than she had in years. The soaring Himalayas thrilled her, but the boulder-strewn landscape of the vast plateau she found on their northern side filled her with solemn joy. "It was CENTRAL ASIA," she later wrote.

On her way home, Bishop met Herbert Sawyer, a young major in the Indian Army. She expressed an interest in Persia, but that country was difficult for foreigners to enter. Sawyer offered to escort her there; apparently he was planning to do a little surveying—or spying—in southwestern Persia. The trip was a terrible one. They rode through blizzards so bitter that Bishop once froze to her saddle, and they nearly starved. To make matters worse, Bishop and Sawyer did not get along. She was glad to see the last of him when his work ended.

Bishop always preferred solitary travel, and her spirits improved as soon as she was on her own again. She spent the next six months riding with a caravan for about 1,000 miles (1,600 kilometers) through Kurdistan, the home of the Kurdish tribespeople in the mountains

John Bishop, the doctor who married Isabella in 1881 after a long and patient courtship.

of western Persia. Then she made her way to the Black Sea, and from there she was able to sail for England. Her *Journeys in Persia and Kurdistan,* in two volumes, was published in 1891.

Bishop had been popular with the reading public ever since *A Lady's Life in the Rocky Mountains* appeared in 1879. By the time her book on Persia came out, she had also won renown as a serious explorer whose reports on out-of-the-way places contained as much information as entertainment. She received an impressive mark of recognition in 1892, when the Royal Geographical Society (RGS) invited her to speak at one of its London meetings. She was the first woman ever to address this exclusive group, and she wrote to her

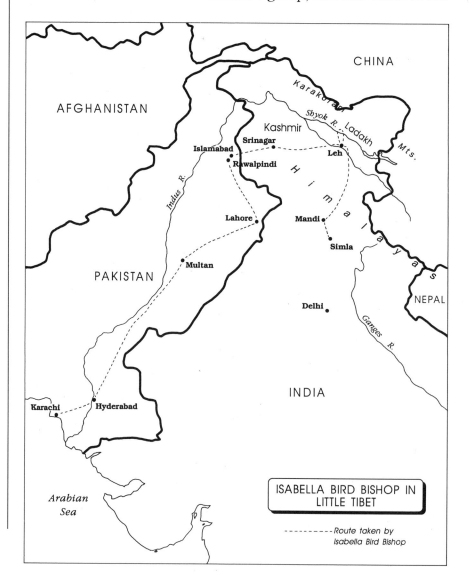

ISABELLA BIRD BISHOP IN
LITTLE TIBET

- - - - - - - - Route taken by
Isabella Bird Bishop

This tent was Bishop's home during the months she spent traveling with the nomadic tribespeople of western Persia. Her book on Persia earned her recognition as a serious geographer, not just a colorful globetrotter.

publisher, "I am grateful for the innovation they have made in recognizing a woman's work." That same year the RGS elected Bishop and fourteen other women to membership. Bishop was the first of these, a groundbreaker in an organization that had always excluded women.

Bishop made two more long trips. She spent the years 1894-97 traveling in East Asia, going by boat far up the Yangtze River and then continuing through southwestern China all the way to the border of Tibet.

During this trip she had the only truly terrifying experience of all her travels. After years of European and American meddling in China's affairs, many Chinese had become hostile to foreigners. Bishop was warned not to go into the remote districts of western China. The people there were extremely rebellious, and she would be far removed from the protection of foreign officials and soldiers. She went anyway, and on several occasions she was chased and stoned by angry crowds screaming at the "foreign devil." Once she was knocked down by a stone and received a concussion ("I felt a mortifying inclination to cry," she admitted). Another time she had to hide from a mob in a darkened room in an inn; she sat for hours with her revolver in her hand, ready to shoot anyone who came through the door, until the crowd broke up and drifted away.

Nevertheless, she continued as she had planned all the way to the foothills of Tibet, where she had a "jolly" time camping with nomads and riding through blizzards with a caravan.

In 1900, at the age of sixty-nine, Bishop wrote, "It is very odd to look at all things in the light of old age, and I am trying resolutely to face it, thankful all the time that my best-beloved never knew it and that they had neither to live nor die alone." Queen Victoria died the following year, and Bishop, like everyone else in the world, felt that an era was passing. But although she was now seventy years old, Bishop found herself not yet ready to give in calmly to old age. She left for North Africa and a 1,000-mile (1,600-kilometer) ride across the Atlas Mountains of Morocco. It was her last trip. She died three years later in Edinburgh, Scotland.

Isabella Bird Bishop was a strange combination—weak and ill when she was surrounded by the comforts of home, tireless and strong when she was traveling in circumstances that most people would have regarded as almost impossibly difficult. An article in the *Edinburgh Medical Journal* after her death said, "When she took the stage as a pioneer and traveler, she laughed at fatigue, she was indifferent to the terrors of danger, she was careless of what a day might bring forth in the matter of food," but at home "she immediately

When she was in her sixties, Bishop spent three years in China. She had mastered the art of photography and managed to take pictures even under primitive conditions, sometimes developing her prints in the muddy water of the Yangtze River. Here she prepares to photograph a group of Chinese townspeople.

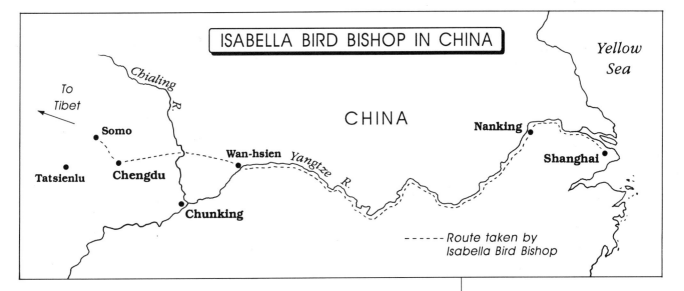

became the invalid, the timorous, delicate, gentle-voiced woman that we associate with the Mrs. Bishop of Edinburgh."

Perhaps Bishop never fully realized it, but she resented the limits that were placed upon her at home. The unspoken rules of English society called for her to be a dutiful daughter, wife, or mother—or, if she wanted more out of life, to do good works in a church, school, or hospital. But something within her rebelled against these conventions and made her ill whenever she tried to follow them. Traveling allowed her to leave them behind, to be happy and healthy. "Travellers are privileged to do the most improper things," she once said, and as soon as she began traveling she took on what she called "an up-to-anything free-legged air." She could wear trousers, fall in love with a desperado, or ride a yak across the Himalayas.

In *The Golden Chersonese,* she described how in Singapore she was offered the chance to visit Malaya. "I was only allowed five minutes for decision," she said, "but I have no difficulty in making up my mind when an escape from civilisation is possible." Travel for Isabella Bird Bishop was never just a trip *to* someplace else— it was an escape, a journey *away* from a life that was too small to hold her.

CHAPTER 3

Florence Baker: Adventures on the Nile

One of the most popular heroines of the mid-nine-teenth century was a brave and resourceful traveler named Florence Baker. With her explorer husband, Samuel Baker, she made two long and extremely hazardous journeys south along the Nile River into uncharted Central Africa. At that time, the search for the sources of the Nile was the world's grandest geographical adventure, and Florence Baker was the first white woman to take part in it.

Samuel Baker's books about the Bakers' expeditions were read by thousands of people who took Florence Baker to their hearts. The public was thrilled to learn how she had taken part in elephant hunts, survived agonizing bouts of malaria, and helped her husband fight off the army of an enraged African king. Yet the admiring public never knew the whole story of Florence Baker's life—a story that was considered highly scandalous by the few people who knew the details. Queen Victoria, who prided herself on her strict moral values, made Samuel Baker a Knight of the British Empire in honor of his discoveries in Africa, but she primly refused even to meet Sir Samuel's wife, his partner in exploration. Not until decades after Florence Baker's death was the full story of her life made public. Fortunately, the twentieth century was more compassionate toward Florence's early history than the nineteenth had been, and she came to be regarded with sympathy and respect as well as with admiration.

Florence Baker, the Hungarian slave who became the wife of an English nobleman, was one of the first Europeans to explore the hidden sources of the Nile River.

Although Samuel Baker was an Englishman, Florence was an Eastern European. She was born in 1841 somewhere in the mountainous region called Transylvania. Her exact birthplace is unknown. Today part of Romania, Transylvania then belonged to Hungary. Florence's family probably lived in one of the German communities that were scattered throughout southern Hungary, for she spoke German as well as Hungarian. She either forgot the details of her background or kept them secret, but she did know that her family was Catholic, that her name was Florence Barbara Maria, and that her family name was Sass. Most biographers have called her Florence von Sass, although her family has never been traced.

Florence lived during a troubled time in Eastern Europe. For centuries the lands that are now Poland, Hungary, Romania, Yugoslavia, and Bulgaria had been a battleground for conflict between the Christian nations of Europe and the Ottoman Empire, an Islamic confederation that ruled the eastern Mediterranean region and North Africa from its base in Turkey. The border of the Ottoman Empire ran through Eastern Europe. That border swung first east and then west as the Europeans, and then the Ottomans, gained the upper hand. Sometimes the population of whole districts was uprooted when the border shifted, and often people fled to escape advancing armies. There were many refugees and people whose nationality was uncertain.

By the time of Florence's birth, the Ottoman Turks had formed an alliance with Britain and France against their common enemy, Russia. The Turks and the Europeans were no longer at war, but civil wars and rebellions broke out all over Europe in the late 1840s. Some historians have called 1848 Europe's Year of Revolution because there were riots and upris-

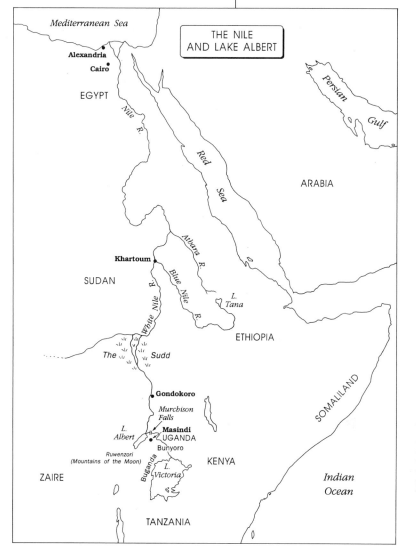

THE NILE
AND LAKE ALBERT

ings in many countries. Hungary was particularly hard hit—fighting erupted everywhere and confusion was widespread.

Florence's family was killed during one of the political or ethnic conflicts that wracked Hungary. Her only memory of the event was of "shots, knives, yells, corpses and fire," she said years later. She was seven years old. Someone managed to hide her during the massacre of her relatives, and she passed into the care of some family servants. Their name was Finnian; she adopted it as her own. Then she vanished into the turmoil that was Eastern Europe after 1848.

Nothing further is known of her until 1859, when Samuel Baker encountered her in Widdin, a town on the Danube River in present-day Bulgaria. Baker was a thirty-eight-year-old widower, with several young daughters who had been cared for by his sister since his wife's death several years before. He came from a wealthy family and did not have to work, so he had devoted himself to adventure. He had lived on several Indian Ocean islands, and he had also spent some time in Constantinople, the capital of the Ottoman Empire. His greatest passions were hunting and shooting, but he also yearned to make a name for himself in African exploration.

The search was underway for the sources of the Nile, and the public's imagination had been captured by the race to explore and map the heart of Africa. Baker felt frustrated that he was not taking part in it. He wanted to accompany the well-known Scottish explorer David Livingstone on an expedition into Central Africa, but Livingstone had turned him down, saying scornfully that Baker did not know how to do anything except hunt. To make up for this disappointment Baker decided to take a trip down the Danube River, which runs through Eastern Europe to empty into the Black Sea.

The Danube carried Baker and his companions east into territory held by the Ottoman Turks. One of their stops was Widdin—a squalid, cheerless town that was a Turkish fortress in Bulgaria. They arrived at Widdin in early 1859, and their Turkish hosts invited them to attend a slave auction.

Slavery was a matter of great and growing concern in the middle of the nineteenth century. The United States was poised on the brink of the Civil War, fought in part over the issue of slavery. In Britain and elsewhere, antislavery societies were springing up to encourage the end of slavery around the world. Slavery was also closely connected with African exploration—one motive of Livingstone and other explorers was the desire to destroy the African slave trade, which was carried on by the Portuguese, Arabs, and Africans.

Baker decided to attend the slave auction in Widdin. Perhaps he had seen such things before, during his earlier visit to Constantinople, for the institution of slavery was as old as history in many parts of the Ottoman Empire. But in Widdin he was shocked to see, among the human merchandise being paraded for sale, a frightened, white-faced girl with fair hair gathered into a braid on her neck.

The aspect of Turkish slavery that most horrified the people of Britain and Western Europe was that people like themselves—white Christians taken by the Ottomans from Eastern Europe and southern Russia—were bought and sold as slaves not only in Turkey but also in the Middle Eastern and North African countries that owed allegiance to the Ottoman Empire. Girls and young women were particularly prized, for Islamic law allowed men who could afford it to keep harems, or establishments of female slaves and concubines. The Ottomans considered it a mark of status for a government official or a wealthy merchant to have a white Christian girl in his harem.

The girl Baker saw at the slave mart in Widdin was seventeen years old. He knew that she was destined for a lifetime behind the walls of an Ottoman harem. Suddenly, to the astonishment of all present, he began to bid for her. When the bidding stopped, she was his.

In later life Samuel Baker never spoke of that day, and neither did Florence, the girl he had bought. No one ever knew what he had to pay for her, or what they said to each other. He was able to speak with her in German because he had spent some time in Germany as a young man. She may have confided the details of her history to him, but no one else ever learned what had become of her after her family was killed in 1848, or how she had fallen into the hands of the slave merchants. All she ever said about her narrow escape from a life of slavery was, "I owe everything to Sam." Baker did not realize it at first, but he had done more than rescue a helpless young woman from degradation. He had found the love of his life and a partner who would be at his side through years of adventure.

Leaving Widdin, Baker and Florence went to the town of Constanza on the Black Sea coast. They spent some months there while Baker decided what to do. He felt he could not take Florence back to England and present her to his family; for one thing, he was not yet sure of his own feelings about her, and marriage seemed out of the question. Yet he was unwilling to leave her, both because she needed his protection and because he was falling in love with her, and she with him.

At this point Baker turned again to his dream of African exploration. He knew that a British explorer named John Speke had gone into East Africa from the Indian Ocean coast and was trying to work his way through unknown country to Central Africa, where he hoped to locate the headwaters of the Nile River. Baker got the notion of going in the opposite direction, working his way from the mouth of the Nile on the Mediterranean Sea up the river toward its source, and perhaps meeting Speke along the way. When he asked Florence if she would care to go up the Nile with him, she said that she would go anywhere he went—even though no European woman had yet braved the dangers of disease, wild animals, and hostile tribes that explorers faced in Central Africa. So they went to Cairo, Egypt, the starting point for all travel on the Nile, and set off on the journey upriver on April 15, 1861.

At this point Florence began to come into her own as a competent, strong-willed explorer. Baker taught her to shoot, and she became a good shot with a lightweight rifle he had used as a boy. When they left the river to take a shortcut across a stretch of desert where temperatures reached 110 degrees Fahrenheit in the shade, she uncomplainingly endured long days in the uncomfortable saddle of her camel, although she was ill from exhaustion and the heat. During half a year or so spent exploring the Atbara River region in the foothills of Ethiopia, she not only organized and managed their camp but also went elephant hunting with Baker. She faced wild animals with calm courage. When she woke up one night to find a large hyena in their tent, she quietly tugged at Baker's sleeve, and he shot the hungry beast. And according to another Englishman who was traveling in the upper Nile districts at the time, she once brought down a charging rhinoceros with a single shot to save Baker's life when his own gun failed.

Baker made the acquaintance of some Ethiopian *aggageers,* warriors who ventured into the wildest regions to hunt, using only

Sir Samuel Baker was a big-game hunter who sought fame as an explorer at a time when Europeans were racing to explore Central Africa. He and Florence were the first white people to see Lake Albert, between present-day Uganda and Zaire.

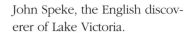

John Speke, the English discoverer of Lake Victoria.

swords and spears against big game such as elephants and buffaloes. The sportsman in Baker greatly admired the *aggageers'* courage and skill. He hunted with them several times, and after he had won their confidence they took him and Florence on long horseback journeys into remote plains and valleys where no Europeans had ever been. Baker and Florence combined hunting with scientific pursuits on these excursions. They made maps of rivers and mountain ranges and took notes about geography, people, and wildlife.

Leaving Ethiopia, Baker and Florence went to Khartoum, a town situated where the two main branches of the Nile River join. Today Khartoum is the capital of the nation of Sudan. In the 1860s it was an Ottoman outpost from which the Turkish rulers of Egypt loosely controlled the Sudan—a vast, largely unexplored region of rivers, swamps, and forests that stretched away south from Khartoum with the Nile at its heart. Baker planned to plunge into this vague immensity with Florence at his side.

First, however, they spent more than half a year at Khartoum, resting and assembling supplies and servants for the trip. Khartoum was a major center of the trade in African slaves, and there Florence and Baker witnessed firsthand the cruelty of the slave dealers and the suffering of their victims. Florence's own experiences had made her detest the slave trade, and now Baker, whose family had kept slaves on its plantation in Jamaica, discovered in himself a new loathing for slavery. Still, they needed the permission of the Ottomans to proceed with their expedition, so they tried to remain on good terms with the governor of the region and other Ottoman officials.

Although Khartoum was 1,500 miles (2,400 kilometers) upriver from the Mediterranean, some European visitors did manage to reach it. In 1862 there were even some women travelers there. A Dutch heiress named Alexine Tinné and her mother, Harriet, were in town when Baker and Florence arrived, and Harriet Tinné's diary gives a portrait of Florence: "A famous English couple have arrived. Samuel and Florence Baker are going up the Nile to find Speke. They have been traveling in Ethiopia and I hear she has shot an elephant!! She wears trousers and gaiters and a belt and a blouse. She goes everywhere he goes." Sam and Florence were not yet married, but he introduced her as his wife.

Baker hired three boats to carry his expedition another 1,000 miles (1,600 kilometers) up the river to the remote, seldom-visited town of Gondokoro, the last place that could be reached by boat along the Nile. South of Gondokoro the river was filled with rapids and waterfalls that no boat could pass. Gondokoro was the edge of the known world; when she left it, Florence would be headed into country that no white woman—and only one or two white men—had ever seen.

The voyage to Gondokoro took them through the Sudd, a huge swamp that ran for hundreds of miles along both banks of the river. The ever-present plagues of mosquitoes were a torment. Both Sam and Florence came down with malaria, a tropical disease carried by mosquitoes. It brings fever, extreme fatigue, and sometimes death to sufferers. Fortunately, they had brought a large supply of quinine, the only medicine that could control malaria.

Gondokoro, when they finally reached it, was no haven of rest. It was both filthy and dangerous. Like Khartoum, it was a center of the slave trade, and it was even farther from the reach of the law. Everyone but the slaves carried guns, and drunken fights and shooting matches broke out often.

Baker ran into difficulties when he tried to move south into the unmapped regions where he believed that Speke was still wandering in search of the Nile. The Arabs he had hired in Khartoum to carry his supplies grew mutinous and demanded to go on cattle and slave raids in the countryside. Baker forbade them to do so, and forty of them attacked him. Florence broke up that menacing situation with the help of the few servants who remained loyal, but Baker confessed later that he felt uneasy about the fate of the expedition from that day forward. Florence remained calm and cheerful, doing her best to smooth over arguments between Baker and the men. But before she and Baker could manage to leave Gondokoro for the south, Speke himself suddenly appeared on the scene with James Grant, his fellow explorer.

Florence and Sam were overjoyed to find that Speke was still alive. He and Grant were the first Europeans to reach the Nile River from the south, and the first to cross Central Africa from south to north. They spent hours talking to Baker and Florence about their adventures in the African kingdoms of Buganda and Bunyoro, of which no European had yet heard (both kingdoms were located in the country that is now Uganda). But Baker was a little crestfallen that

Speke had succeeded in exploring Lake Victoria, the principal source of the Nile. He wondered if there was anything left for him to discover.

He cheered up when Speke suggested that he should look for a large lake said to lie somewhere west of Lake Victoria, with high mountains on the far side of it. Speke and Grant had heard rumors of this lake, which the natives called Luta N'Zige (Dead Locust), but Kamrasi, the king of Bunyoro, had prevented them from searching for it. Speke thought that Luta N'Zige might be another source of the Nile. Baker's imagination was fired at once. He would seek this mysterious lake; perhaps it would be a discovery even more important than Lake Victoria.

Such a venture, however, was sure to bring unpredictable hazards: disease, possible shortages of food or water in unexplored country, hostile natives (Speke and Grant warned Baker that Kamrasi was hostile and treacherous), and perhaps betrayal by Baker's own men. Speke and Grant were appalled that Baker would even consider taking Florence along on such a journey.

Soon after her arrival in Africa, Florence proved her toughness by surviving a scorching desert crossing. Later she was to face still greater hazards.

They planned to sail downriver to Khartoum and Cairo in Baker's boats, and then home to England; they could easily have escorted Florence to someplace safe to await Baker's return. Sam would certainly have sent Florence downriver to safety if she had wanted to go. It seems most likely that she simply did not wish to leave him.

Baker and Florence left Gondokoro several months later, bound for the kingdom of Bunyoro and the lakes of Central Africa. The march started well enough, but soon became hellish. The horses, camels, and donkeys began to die of sleeping sickness carried by tsetse flies. The few men whom they had persuaded to come with them were mutinous and unreliable. Florence and Sam took turns keeping watch at night to avoid being killed, but they could not prevent the men from simply leaving the expedition, and some of the deserters pilfered from their fast-dwindling supplies. The travelers ran out of quinine; malaria and other illnesses began to prey upon them. Both Sam and Florence became so weak that they sometimes had to be carried in litters called *angareps,* something like portable bed-frames.

But they kept going. By early 1864 they had reached a point farther south than any Europeans had yet reached from the north. Their route was slightly different from Speke's northward march, so they were passing through an utterly unexplored region. It consisted of flat brushy plains interrupted by many marshes and swamps. Baker tried to make geographic observations whenever he could, but he was hindered by his increasing weakness.

Finally they reached Kamrasi's kingdom. The king asked for Florence as a gift, whereupon the outraged Baker threatened to shoot Kamrasi. Despite this hostile exchange, the king gave them permission to pass through Bunyoro on their way to look for Luta N'Zige.

Florence almost died during the next leg of the search for the lake. While crossing a swamp on a narrow bridge of reeds, she suffered some sort of seizure, probably brought on by heat and exhaustion. She fainted and did not regain consciousness. Baker walked beside her *angarep,* holding her head up so that she would not choke, and dripping water into her mouth to keep her alive. Days later she woke from her

When Speke and Grant walked out of unknown Central Africa at Gondokoro, they found the Bakers in residence there. Here Baker, at far right, listens intently to Speke's tale of a mysterious lake waiting to be discovered.

coma to the sound of hoes and shovels—Baker, afraid that she was dying, had despairingly ordered the men to dig her grave.

Several weeks later, in March 1864, Baker and Florence climbed the crest of a hill. Ahead of them, shimmering in the distance, they saw a broad gleam of silver. It was Luta N'Zige—at last they had reached their goal. They made their way down a twisting path to the shore as quickly as they could. On the lakeshore, each made a gesture to commemorate their hard-won achievement. Baker recalled, "I rushed into the lake, and thirsty with heat and fatigue, with a heart full of gratitude, I drank deeply from the Sources of the Nile." It had not yet been proved that the Nile issued from Luta N'Zige, but Baker was certain that it did. He also knew that if he could only return to London with word of this vast new lake, he would become as famous as Livingstone and Speke.

Florence's gesture was more nostalgic. Although she never talked about her childhood and had no desire to return to her homeland, she did wear a hair ribbon in the colors of Hungary's flag: green, red, and white. Still exhausted from her recent illness, she hobbled to the water's edge and tied the ribbon to a bush. In this way Hungary and Britain shared the credit for discovering the lake, which Baker dubbed Lake Albert in honor of Queen Victoria's husband.

The expedition now took to the water and headed north on Lake Albert in two dugout canoes bought from local villagers. At one point Baker heard of a large waterfall not far from the lake. He wanted to try to find it, but Florence was ill again, so he asked her what they should do. "Seeing is believing," she told him, declaring that they should locate the waterfall before turning back to Gondokoro. A true explorer despite all she had suffered during her years in Africa, she shared Baker's passionate desire to fill in more of the blank space on the map.

They headed up a river that flowed into the lake. Tall cliffs rose on either side of the deep, rushing water. Before long they heard a roaring sound like thunder, but the sky was clear. Then, around a turn in the river, they came upon a sparkling cataract that plunged down a high rock wall into a foaming pool. They were the first explorers to see this waterfall, the largest one on the Nile; it was a discovery almost as momentous as Lake Albert. Baker named the cataract Murchison Falls after Sir Roderick Murchison, the president of the Royal Geographical Society. Now all that remained was to tell the world what he and Florence had found.

The trek north to Gondokoro was grim. Kamrasi detained them for months, hoping that they could help him win a battle against a rival African chieftain. Food supplies were dangerously low, and both Florence and Baker suffered cruelly from malaria; often they were too weak to get out of bed for weeks at a time. Finally they made their way to Gondokoro with a caravan of ivory and slave traders. They reached the town in March 1865, a year after sighting Lake Albert.

At once they hired a boat and sailed down the Nile. By September they had reached the Mediterranean Sea. They arrived in London less than a month later, and there they were married.

Samuel Baker's new wife caused almost as great a sensation as their African discoveries. Many people wondered where he had met her and when he had married her. Although the details of the slave auction in Widdin remained a secret, speculation about Florence's background was enough to make some women—including Queen Victoria and some of Baker's stuffier relatives—refuse to have anything to do with her because they thought that she was somehow immoral. Yet Florence also made many friends among Baker's acquaintances and family, and she was a diligent and affectionate stepmother to his daughters.

To the public at large, she was simply a marvelous heroine. When Baker began writing a book about the expedition, he and his publisher quickly realized that readers would be fascinated by the story of the first white woman in Central Africa. Baker was careful to include her in his account. Writing of the triumphant moment when they approached the Mediterranean on their homeward journey, he said, "Had I really come from the Nile Sources? It was no dream. A witness sat before me; a face still young, but bronzed like an Arab with years of exposure to a burning sun; haggard and worn with toil and sickness and shaded with cares, happily now passed; the devoted companion of my pilgrimage to whom I owed success and life—my wife."

Florence Baker was widely regarded as a model of pluck, courage, and strength. The Bakers were presented with medals by geographic societies in Paris and elsewhere. Sometimes, on these occasions, Florence received more applause than Sam.

Sir Samuel and Lady Florence Baker made a second journey to Africa in 1870-73, again traveling up the Nile and overland to Lake Albert. This time they started with a large, well-equipped force. Their plan, supported by the ruler of Egypt under pressure from the British government, was to claim Bunyoro and the rest of the lake region for Egypt. They also intended to outlaw the slave trade in the Sudan.

Florence performed a scientific role, keeping records of the weather and gathering samples of plants. She also kept a diary in English, which she had taken pains to learn; parts of Florence's diary were published in 1972 under the title *Morning Star* (the Africans called her Njinyeri, which means "morning star"). When they reached Bunyoro, Florence wrote, "I am sorry to say that this country reminds me of great misery. When I see all the old faces, then I cannot help thinking how we both suffered from illness and misery."

Ultimately, the Bakers' second expedition was a failure. Kabarega, the new king of Bunyoro, imprisoned them and their followers in his capital city, Masindi. They had to fight their way out of Masindi and flee back to Gondokoro. Florence later wrote to one of Sam's daughters, "My darling child, it is quite impossible to tell you about our weary march—I can only tell you that the entire population lay

Florence Baker, dressed like a well-bred Victorian lady in London, confronts an African village. On her first African journey she wore men's trousers and high boots—much more practical for exploring than a dress. She was sometimes mistaken for Baker's son.

in ambuscades, and we had to fight for seven days through that dreadful country, where it was quite impossible to see the enemy...only showers of spears passed our faces."

But the Bakers survived and returned to England, where they settled down in a mansion called Sandford Orleigh. Baker wrote several more books about Africa, but he never traveled there again— Florence refused to go, and he would not go without her. They did, however, travel to the American Rockies, to India, and to Japan. Sir Samuel Baker died in 1893. Lady Florence Baker lived on until 1916, keeping Sandford Orleigh exactly as it had been during Sam's lifetime. She died in the middle of World War I, the last survivor of that hardy group of nineteenth-century explorers who solved the age-old mysteries of the Nile.

CHAPTER 4

Fanny Bullock Workman: Himalayan Heroine

T he two best-known American world travelers of the late nineteenth and early twentieth centuries were a married couple: Fanny Bullock Workman and her husband, William Hunter Workman. They traveled almost nonstop, in devoted harmony and complete equality, for twenty-five years. During that time Fanny Bullock Workman became the world's foremost woman mountaineer, and in 1906, at the age of forty-seven, she set a mountain-climbing record for women that was not broken until 1934. She was an outspoken feminist and outdoorswoman—and also one of the most controversial figures in modern geography.

Fanny Bullock was born in Worcester, Massachusetts, in 1859. Her family was prosperous and socially prominent; her father was elected governor of Massachusetts when she was seven. She was educated by private tutors and at elite schools in France and Germany. In 1881, when she was twenty-two years old, she married thirty-four-year-old Dr. William Hunter Workman, a successful physician. They had one child, a daughter named Rachel.

In 1889 Dr. Workman decided to give up his medical practice because his health was poor. He did not need to work; the Workmans

Fanny Bullock Workman and her husband, William Hunter Workman, were well matched: They were two of the hardiest and most persistent travelers the world has ever seen.

had plenty of money. They moved to Germany. There Dr. Workman's health must have improved dramatically, for soon he and his wife had taken up a vigorous outdoor life, hiking in Norway and Sweden and climbing in the Alps. Fanny Bullock Workman was one of the first few women to climb the Matterhorn peak in Switzerland. She also climbed Mont Blanc, Europe's highest mountain at 15,781 feet (4,813 meters).

The Workmans enjoyed their free and easygoing excursions. By 1895 they were ready to give up conventional life altogether and devote themselves to travel. They deposited Rachel in boarding school and set off on a long bicycle trip that carried them through Algeria, in North Africa, and Spain. With blankets, teakettles, water bottles, and tire-repair kits strapped to their bicycles, they found that they could travel almost anywhere. They brandished revolvers to scare off bandits and whips to fight off wild dogs. Up mountains, across rivers—they pedaled when they could and pushed when they had to. Both of them developed great stamina. And they kept methodical records of every imaginable detail: how many miles they covered each day, how many flat tires they mended, what routes they

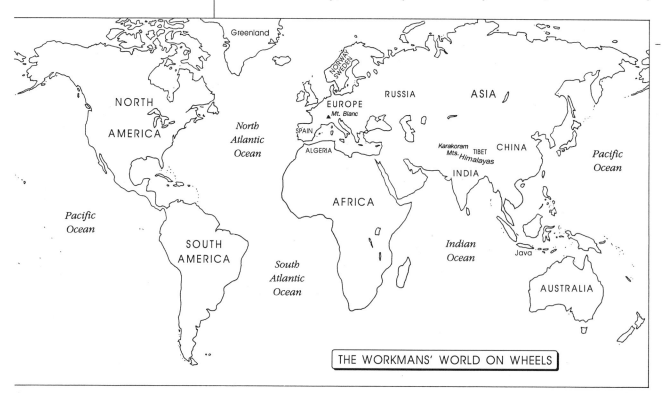

THE WORKMANS' WORLD ON WHEELS

followed. Later they turned these notes into popular books that other cyclists used as tour guides. In writing, as in traveling, they made a good team.

During these years the issue of women's rights began to be more and more discussed in Europe and especially in the United States, where women were fighting for the right to vote. The Workmans believed strongly in women's equality with men, and Fanny in particular took every opportunity to speak up in favor of women's rights. She felt that her own life proved that women were capable of being men's equals. Throughout her wide-ranging travels, she grew indignant whenever she saw women being treated as second-class citizens or—even worse—as mere possessions, and she wrote often of the need for more freedom for women in Africa and Asia.

But although she was a feminist, there was one area in which she remained old-fashioned and conventional. Some women had begun to wear divided skirts, or even full-legged trousers, for bicycling and outdoor activities, but Fanny Bullock Workman would not hear of such a thing. She always wore a skirt (and a hat with a veil). At first her skirts were so long that they dragged on the ground and tangled in her bicycle wheels; later she wore slightly shorter skirts and modestly wrapped her legs in thick cloth padding.

In 1897 the Workmans were ready for a bigger challenge than the Mediterranean world could offer. Loading their bicycles, revolvers, and notebooks onto a steamship, they headed east. They planned to cycle across the entire subcontinent of India, from north to south and east to west. But the idea of traveling for simple pleasure seemed frivolous to them. They needed a serious purpose for their trip, so they made an intensive study of the art and architecture of India's Hindu religion and then decided to see and photograph as many ancient Hindu palaces and temples as possible. India abounds in ruined palaces and temples. The Workmans eventually pedaled across 14,000 miles (22,400 kilometers) of desert, jungle, and plain—with side trips to Hindu and Buddhist ruins in Burma, Java, Ceylon (now known as Sri Lanka), and Cambodia.

They bowled along through the heat, dust, and monsoon rains of India as steadily as if they had been riding down a country lane in New England. Although India is a huge and endlessly varied land with hundreds of different languages and regional cultures, the

At a place called the Silver Throne Plateau, more than 20,000 feet (6,557 meters) above sea level on Pakistan's Siachen Glacier, Fanny Workman holds up a sign reading "Votes for Women." Feminism had reached the ends of the earth.

A bridge over the Indus River. Such bridges, made of woven grass and tree branches, were almost as perilous as the peaks and glaciers of the Himalayas.

Workmans paid less attention to the people around them than to the ruins and the scenery. They prided themselves on being tough and resourceful and on their indomitable energy and health. India was then under British rule, and Fanny wrote that the British people in India lived in continual fear of "three bugaboos": fever, sunstroke, and catching a chill. She added rather smugly, "Without discussing these here we may say briefly that, in spite of constant exposure to heat, sun, cold, wet, and malarial emanations, in the course of many thousand miles of travel in all parts of India, we escaped all of these evils."

The India project took the Workmans more than three years and resulted in a successful book called *Through Town and Jungle,* but in the end it was just a warm-up for their most ambitious work, the exploration of the Himalaya Mountains.

The Himalayas are the world's highest mountain range. They run in a long arc across the top of India, dividing the Indian subcontinent from the plateau of Tibet and the highlands of Central Asia that today are divided among Afghanistan, the Russian commonwealth, and China. Although the entire range is sometimes called the Himalayas, several other mountain ranges are included in its western part. Chief among these are the Karakoram Mountains, which meet the Himalayas in a massive knot of peaks and glaciers high above the plains of northwest India and present-day Pakistan. This mountain complex was the last frontier of exploration in India. Native rulers had closed much of it to outsiders for centuries, and great stretches of it were still unexplored at the end of the nineteenth century.

In 1898, during a break in their bicycle tour, the Workmans visited Kashmir, a province in northwestern India long renowned for the beauty of its landscapes, especially its tranquil lakes set against a

The climbers creep up a wall of ice on Mt. Lungma in the Karakoram Mountains. Fanny reached the top of this mountain and set a new altitude record for women.

backdrop of green terraced hills and snowy mountains. From Kashmir they traveled by foot and pony along a trail through the western Himalayas and the Karakorams. They were seized with a passion to climb these magnificent peaks.

Mountaineering was then a relatively young sport. People did not begin climbing peaks in the European Alps for recreation until the late eighteenth century. When the Workmans arrived in the Himalayas, sport climbing had just been introduced to Asia by the British. W. W. Graham had made the first recorded climb in the Himalayas "purely for sport and adventure," as he said, in 1883. Other climbers followed, and many of them, encouraged by the Royal Geographical Society and the Survey of India, added a little mapmaking and some scientific observations to their climbs. The Workmans, who had enjoyed climbing in Europe, decided to devote themselves to Asia's grand mountains. They put away their bicycles and sent for warm tents and sleeping bags.

Very little climbing gear was available at that time. There were no oxygen bottles, no lightweight nylon ropes, no miracle fabrics to keep climbers dry and warm. The Workmans made their own climbing boots by driving steel nails through the soles of their walking shoes. They wore wool, which soon grew damp and heavy, and their only tools were ice axes and ropes.

Although she usually bicycled or walked in India, Fanny occasionally agreed to be carried in a portable chair called a *dandi*.

The Workmans had plenty of climbing experience, but they soon learned that climbing in the Himalayas was quite different from climbing in the Swiss Alps. In Switzerland, even the highest peaks were surrounded by villages and railways, and huts had been built on the mountainsides to store supplies. A climber was rarely more than a day's walk away from civilization. The Himalayas, however, were a vast, inhospitable, and largely uninhabited world. Often a trek of many weeks was required just to reach the base of a mountain. All supplies had to be carried in over great distances by native porters. The organization and management of a Himalayan expedition was sometimes more difficult than the actual climb. Several of the Workmans' mountain expeditions failed, in fact, because they had trouble with their porters. On one occasion relations between the Workmans and their servants grew so bad that 130 porters deserted them in the night, taking much of the food with them.

The Workmans made one attempt to climb in the eastern Himalayas. In the distance they could see Mt. Everest, the world's tallest peak, "floating up from a mellow haze to an unappreciable height," as Fanny rapturously described it. But they had to turn back when their forty-five porters refused to go on through heavy snow. Thereafter they moved their operations to the western Himalayas and the Karakorams. Between 1899 and 1912 they made six climbing and exploring expeditions into this remote world of rock and ice.

On the first of these expeditions Fanny Bullock Workman set a world altitude record for women by climbing Mt. Koser Gunge, which is 21,000 feet (6,384 meters) tall. This was a genuine triumph, something worthy of the world's notice, and Fanny found it intoxicating. From then on the Workmans were driven in part by her desire to set new records and to do things that no other woman had done.

Short, somewhat stout, middle-aged, and swathed in skirts and veils, she slogged determinedly up and down icy slopes, occasionally admitting that she felt exhausted, or cold, or afraid—but she always claimed that by the next day she "felt perfectly fit and able to attack another mountain had it been necessary." She and William took turns managing the expeditions. On one, he would organize the porters and supplies while she, with the advice of imported Swiss guides, would plan the route and the climb; on their next trip, they would switch jobs.

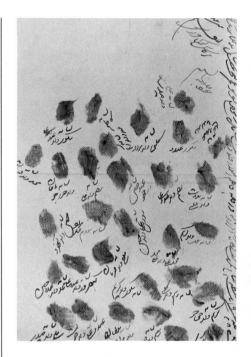

Local porters signed contracts with their thumbprints. The Workmans argued frequently with their porters, who were accustomed to a more leisurely pace than that demanded by the impatient Americans.

A porter carries Fanny across a river at the center of the Siachen Glacier in the eastern Karakorams. The Workmans were the first to explore this ice field and climb the peaks around it.

In 1902 and 1903, the Workmans explored a 30-mile (48-kilometer)-long glacier in a little-known Karakoram region called Baltistan, which is now the northernmost part of Pakistan. There Fanny broke her own altitude record by climbing Mt. Lungma (22,568 feet, or 6,883 meters). William also set a new men's altitude record, reaching 23,392 feet (7,134 meters) on another peak in Baltistan. But although William's achievement was noteworthy, especially for a man of fifty-six, he was always more proud of his wife's accomplishments than of his own. He was an unusual man for his time—happy to accept his wife as his equal in every way. In one of their books he praised her "courage, endurance, and enthusiasm," and he rejoiced that together they had "shared equally all the excitements, hardships, and the dangers of the adventurous life." The editor of the *Alpine Journal* wrote that William Hunter Workman wanted his wife to be respected by other climbers; Workman "was never so keen or quick as in her support," the editor added.

The crown of Fanny Bullock Workman's mountain-climbing career came in 1906, when she was forty-seven years old. The Workmans were exploring a dense, steep cluster of many peaks called the Nun Kun Massif in Kashmir. Fanny wrote, "The rugged and savage beauty of this group, and the evident complexity of its formation, proclaimed it a most alluring field for investigation....And this time we decided to employ Italian porters. For we have had more than enough, in the past, of sitting on cold snow-slopes, awaiting the snail-like approach of unwilling coolies, and hearing their wailing complaints and refusals to march."

The expedition, although costly, ran smoothly, and both Fanny and William reached the top of Pinnacle Peak in the Nun Kun Massif. They determined its height to be 23,000 feet (7,015 meters). Very few men had climbed to this height, and no woman had done so. Fanny had set yet another record.

This record stood for many years. Pinnacle Peak was later measured more accurately and shown to be only 22,815 feet (6,958 meters) tall, but it was still the highest point yet reached by a woman.

Fanny liked to think of herself as a trailblazer for other women—but she did not want them to follow her *too* closely. She was fiercely proud of her world record, so proud that she caused a controversy several years later when another woman claimed to have broken it.

Annie Smith Peck, a rival American pioneer of women's mountaineering, climbed one of the peaks of Mt. Huascaran, the highest mountain in Peru, in 1908. Peck had very little money and her climbing expeditions were equipped with only minimal scientific instruments. Her reports of altitude were based as much on guesswork as on technical measurements. But she claimed that the summit of Huascaran was at least 23,000 feet high, maybe higher. She announced in *Harper's Magazine,* "If, as seems possible, the height is 24,000 feet, I have the honor of breaking the world's record for men as well as for women." Peck repeated her claim in the journal of the American Geographical Society. The result was a feud between Peck and the Workmans in the pages of the world's scientific journals.

Fanny was not prepared to have her record challenged by someone who merely "estimated" altitude. She spent $13,000 to send a team from the French Geographic Service to Peru to conduct a rigorous scientific measurement of Huascaran. The result: Peck's peak was only 21,812 feet (6,652 meters) tall. The Workmans published their report in *Scientific American,* demolishing Peck's claim. Although Peck sarcastically replied that she had not expected anyone

TWO SUMMERS IN THE ICE-WILDS OF EASTERN KARAKORAM
THE EXPLORATION *of* NINETEEN HUNDRED SQUARE MILES OF MOUNTAIN AND GLACIER
By FANNY BULLOCK WORKMAN *and* WILLIAM HUNTER WORKMAN
WITH THREE MAPS AND ONE HUNDRED AND FORTY-ONE ILLUSTRATIONS BY THE AUTHORS

Cloud from avalanche descending between two granite peaks on Bilaphond glacier.   Frontispiece.

E. P. DUTTON & COMPANY
681 FIFTH AVENUE, NEW YORK

Fanny and William took turns as leader, in both their expeditions and the writing of their books, such as this chronicle of their last expeditions in the Karakorams. Fanny once wrote about her conquest of Mt. Koser Gunge, "For the benefit of women, who may not yet have ascended above 16,000 feet but are thinking of attempting to do so, I will here give my experiences for what they are worth." She saw herself as a role model for other women climbers.

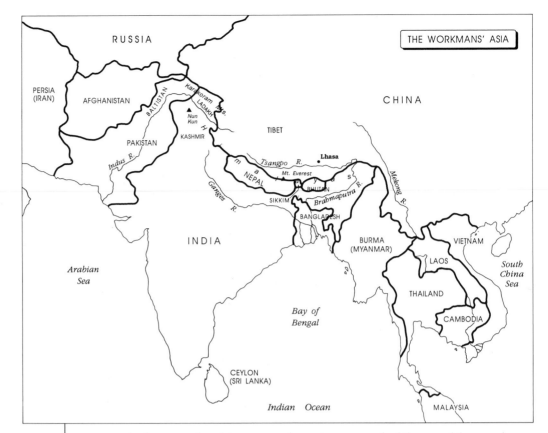

to spend a fortune checking up on her, she conceded the world record to Fanny Bullock Workman. Fanny's record would not be broken for twenty-eight years.

Unlike most mountaineering expeditions, the Workmans' trips were not sponsored by a government survey department or a scientific research organization. They made their own plans and paid their own expenses. But they did not want to be regarded merely as recreational climbers. They carried scientific instruments and cameras, they measured and photographed everything, and they wanted the five books they wrote about their mountaineering expeditions to be seen as serious contributions to geography.

The Workmans received many honors and became members of many scientific and learned societies. Fanny was especially proud of winning recognition that was usually withheld from women. In 1905, she was invited to speak before the Royal Geographical Society (RGS) in London. She was only the second woman to do so since the RGS was founded in 1830; Isabella Bird Bishop had been the first. Fanny Bullock Workman was also the first woman to lecture at the Sorbonne, a renowned college in Paris.

The Workmans were looked down upon, however, by some of their fellow explorers. For one thing, they were Americans. The British had long regarded India—particularly the Himalayas—as their private terrain, and many British experts were inclined to be scornful of American efforts. For another thing, the Workmans were undoubtedly pushy, aggressive, and competitive. They were often accused of being more interested in setting records than adding to knowledge. It is true that some of their claims were not quite accurate (although mountain surveying was and is a difficult science, and the Workmans' errors were almost certainly honest mistakes). Finally, some people refused to take Fanny's achievements seriously because they believed mountaineering should be for men only. But the Workmans' eighth and last expedition—of which Fanny was the leader—won them praise as explorers and geographers even from their critics.

That expedition, in 1912, took them deep into the eastern Karakorams to the 45-mile (72-kilometer)-long Siachen Glacier. Fanny climbed several peaks in the vicinity and explored the glacier thoroughly. She discovered a new pass through the mountains around the glacier and pioneered a route through unknown snow-fields to another glacier. The Workmans wrote about their 1911 and 1912 expeditions in *Two Summers in the Ice Wilds of the Eastern Karakoram: The Exploration of Nineteen Hundred Square Miles of Mountains and Glaciers.*

Published in 1917, this book was the Workmans' masterpiece and the culmination of their careers as explorers. Kenneth Mason, a professor of geography at Oxford University who had sneered at the Workmans' earlier claims to be explorers, admitted that it was "a fine achievement." And Sven Hedin, a Swedish explorer who was internationally recognized as an expert on Central Asia, called it "one of the most important contributions ever given to our knowledge of these mountains."

Fanny and William Workman retired from mountaineering after the Siachen Glacier expedition. They made lecture tours throughout Europe and the United States, and during World War I they lived quietly in southern France. Fanny died there in 1925 at the age of sixty-six. Her will revealed that she had left money to four women's colleges in the United States: Radcliffe, Wellesley, Smith, and Bryn Mawr. William Hunter Workman, whose poor health had brought about their move to Europe years before, outlived his intrepid wife by a dozen years. He died in Massachusetts in 1937.

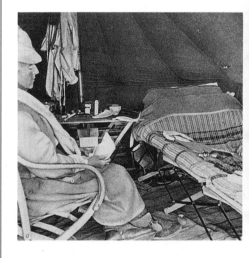

Fanny and her camp tent on the Nun Kun expedition of 1906.

CHAPTER 5

Mary Kingsley: Wandering Through West Africa

Perhaps more than any other woman explorer of her time, Mary Kingsley traveled to get away from a sad, empty life at home. Once she reached the jungles and mangrove swamps of West Africa, however, her love of natural beauty blossomed, and so did her keen sense of humor. Her books about her West African adventures are among the wittiest and most entertaining travel stories ever written, whether she is describing a close encounter with a crocodile or dinner in a cannibal village.

Yet Kingsley was not—as she has sometimes been portrayed—just a funny spinster who canoed through Africa in a long dress. Not only did she make a real contribution to scientific and geographic knowledge, but her views about the African people were the most enlightened and respectful of her day.

Kingsley was born in 1862 in London, England. Her father, George Kingsley, came from a prominent family of writers and clergymen; he had been trained as a doctor. Her mother, Mary Bailey, was one of George Kingsley's servants. They were married four days before Mary was born.

Mary Kingsley traveled alone through West Africa in the late nineteenth century. Unlike most Europeans, she admired and respected the African way of life.

The circumstances of Mary's parents' marriage and her birth were scandalous in mid-nineteenth-century England, and they created a rift between George Kingsley and his brothers. As a result, Mary's contact with her uncles and cousins was limited. She was always aware of the shadow of scorn that hung over her family, although she did not know the truth about her parents' marriage until after they were dead.

George Kingsley and his wife had a second child, Mary's younger brother Charles. Their marriage, however, was apparently not a happy one. Mary's father spent most of his time away from home. He liked to travel and was able to find work as an attending physician to traveling noblemen. He dabbled in science, collecting biological specimens and cultural artifacts from all over the world, but he did not always send his family in London enough money to live comfortably. His wife soon took to her bed as a perpetual invalid, and much of the responsibility for running the house and taking care of Charles fell upon young Mary.

So Mary Kingsley's childhood was fairly dreary. "The whole of my childhood and youth was spent at home, in the house and garden," she later recalled. "I knew nothing of play or such things." She had no opportunity to make friends. She had an unquenchable thirst for knowledge, but she received no formal schooling, although her parents managed to provide Charles with an expensive education. "I cried bitterly at not being taught things," Mary later said.

Somehow she learned to read, however, and then she quietly proceeded to educate herself by raiding her father's library. "No one would believe the number or character of the books I absorbed," she said. "I did not say anything about them, finding if I did, it generally meant an injunction not to do it." Her favorites were books by travelers and explorers and a lurid volume called *Robberies and Murders of the Most Notorious Pirates*.

In the early 1880s, Mary's father gave up his roving life because of poor health. Mary, who adored him, was happy to have him home. He moved his family to Cambridge, England, where Charles was attending college. In Cambridge, for the first time, Mary had something of a social life. In her father's company she met some of the scholars and writers who were associated with the university and she engaged in as many learned discussions as she could. She met other intellectual young women and formed a few close friendships.

Mary also spent a good deal of time helping her father, who was trying to organize a lifetime's worth of notes on travel and science into publishable form. At this time he made his only contribution to her education—he paid for her lessons in German so that she could translate articles from German magazines for him. At the same time she continued to be solely responsible for the maintenance of the household and the care of her mother.

Her responsibilities deepened a few years later when both parents' illnesses worsened. She nursed them attentively until, in 1892, both died. She was almost thirty years old, but even now her life was not her own. As Elspeth Huxley, a twentieth-century Englishwoman who also explored in Africa, says, "Mary Kingsley belonged to an age and generation when women's subjugation was so complete that even the death of both parents did not really set her free." Mary's brother Charles now expected her to look after him, although he was twenty-six years old, and she accepted this as her duty. In her heart, though, she had a new dream—to travel, to see something of the world she had read about and heard other people talk about throughout her shuttered life.

Several months later, Charles departed on a trip of his own—to Asia—and Mary Kingsley was finally free. She too wanted to see Asia, but she knew she could not afford it. But with a little money that her parents had left her, she decided to go to the Canary Islands off the coast of Morocco. For centuries the Canaries had served as a way station for ships and passengers traveling between Europe, Africa, and the Americas. They were familiar enough to be respectable, but close enough to tropical Africa to be exotic—the perfect choice for the first-time traveler.

Kingsley loved the week-long steamship trip to the islands. And her first sight of Tenerife, a high mountain in the Canaries, seemed to her to be an unearthly vision, "an entirely celestial phenomenon." It "stood out a deep purple against a serpent green sky, separated from the brilliant blue ocean by a girdle of pink and gold cumulus [clouds]." To Kingsley, who wore her blond hair pulled back tightly into a bun and had dressed in black since her teens, the world away from home always appeared carefree and vividly colorful.

She wrote to a Cambridge friend, "I have been having a wild time." She made an excursion to see a volcano and spent the night camped in the open air; she also made at least one voyage on a trading boat

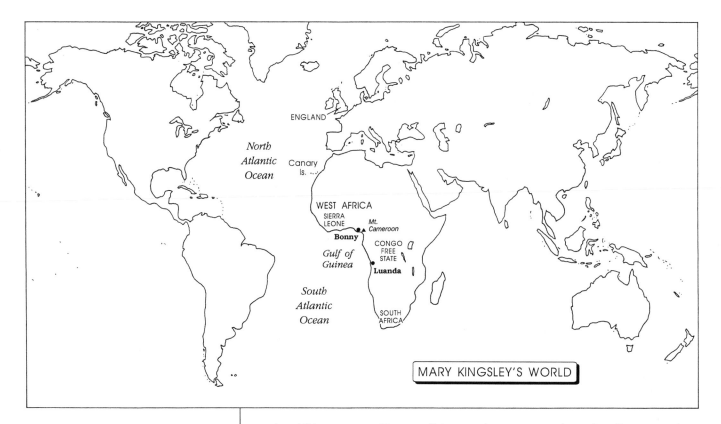

MARY KINGSLEY'S WORLD

to the African coast. From talking to the coast traders she discovered that there were some hardy individuals who traveled alone through the bush country, walking or canoeing from village to village with cloth, knives, and other trade goods that they bartered with the Africans for ivory and rubber. Of course, all of them were men. But Kingsley believed that she could do the same. She returned to London to make plans.

The long, curved coast of West Africa from Sierra Leone to Angola had been carved into districts of influence by Great Britain, France, Germany, Belgium, and Portugal. The colonies were nothing more than settlements on the coast or a little way inland, and political boundaries were not very strict—people could easily travel from one district to another. Steamers plied the coastal waters. But much of the interior had not yet been explored, and travel there was a much more uncertain thing. Furthermore, while many of the coastal African tribes had been pacified by the Europeans, some tribal groups in the interior remained hostile and therefore dangerous to intruders.

It was into this region that Kingsley proposed to go. She was well aware that she might not return. Six years later she wrote to a friend explaining the origin of her 1893 trip: "Dead tired and feeling no one had need of me any more after Mother and Father died within six weeks of each other in '92 and my brother went off to the East, I went down to West Africa to die." She may have been joking when she wrote that letter, or feeling sorry for herself, or both—but she may also simply have expected the trip to kill her.

The area she planned to visit was notorious for deadly diseases. Europeans who went there died in droves from fevers and parasitic infections. Kingsley protected herself as well as she could by drinking only water that had been boiled (to kill bacteria) and by taking quinine pills (to combat malaria), but many other Europeans followed the same precautions and died anyway. She said wryly that on her trip to the Canary Islands she had noticed that whenever she talked to a veteran of the coast trade, his stories would be punctuated by the frequent remark, "He's dead now." She had thought it funny at the time, but after traveling in Africa and seeing the death toll among foreigners there, she understood.

There were other dangers to be considered as well, although Kingsley was too well-bred, by the prim standards of her time and place, to discuss them openly. Some travelers in West Africa had been murdered, or had simply disappeared. The same thing could happen to her. And, like every woman who has ever traveled alone, she was aware that she was especially vulnerable just because she was a woman—she ran the risk of being sexually assaulted. She carried a small revolver, although she never used it. She also kept a small, sharp knife hidden in her clothing, believing that she would use it on herself if she ever found herself in an unbearable situation. Fortunately, this drastic measure never became necessary.

The only white women along the African coast were a few missionaries and wives of colonial administrators. People were shocked and alarmed by the idea of a woman traveling alone in the region—and paying her way by bargaining with the Africans like a common trader! But Kingsley calmly assembled her trade goods and set out. Before long many of the people who had been shocked by her boldness were won over by her courage, intelligence, humor, and charm. She made many friends among the traders and ship captains of the coast, and also among the officials she visited.

She found that being a trader helped her get along with the Africans, especially in regions away from the coast where many of the people she met had never seen a European. Later she explained to a lecture audience that "there is something reasonable about trade to all men, and you see the advantage of it is that when you first appear among people who have never seen anything like you before, they naturally regard you as a devil; but when you want to buy or sell with them, they recognize there is something human and reasonable about you." She developed into a moderately shrewd bargainer and managed to pay for her trips with the rubber and other items she acquired.

Kingsley's first trip to Africa took place in 1893. What actually happened is rather a mystery, for although she took notes she did not publish them, and some of her notes and letters from this period were later lost or destroyed. She went down the coast by sea as far south as the settlement of Luanda in Portuguese Angola (now the independent nation of Angola). From there she worked her way back up the coast to Bonny, in present-day Nigeria, and returned to London by sea.

She saw much that was disturbing on this journey. In the Congo Free State or Belgian Congo (today the country of Zaire) she saw the tragic results of widespread, devastating epidemics of smallpox and sleeping sickness. She also witnessed the atrocities that were being committed there by the Belgian government—the enslavement, abuse, and murder of thousands of Africans. (Six years later Joseph Conrad would describe the horrors of the Congo in the novel *Heart of Darkness*.) In *A Voyager Out,* a biography of Kingsley, Katherine Frank says that Mary was so appalled by these atrocities that she vowed never to enter the Congo again "until it was in different hands."

The trip was not all bad, however; in fact, much of it was very good. Like many women travelers of her era, Kingsley

thought there was something a little too self-indulgent about the idea of traveling just for fun. She needed what she called "odd jobs to do"—something to provide a focus for her curiosity, a serious purpose for her wandering. She came up with "fish and fetish." *Fish* meant the collecting of specimens of river fish for the British Museum, and *fetish* meant the study of native religious beliefs and sacred objects, or fetishes. She came back with many examples of both fish and fetish, and she was especially pleased when an expert at the British Museum praised her fish specimens and offered her the use of professional collecting equipment for her next trip.

For she was already planning her next trip. The first one, it seems, had been an initiation of sorts. She had passed all the tests of travel; she had learned to eat, sleep, and get around in an utterly foreign place, and she had enjoyed it immensely. Now she regarded herself as an experienced African traveler and was eager to go back. England had little to offer her—no family, only a small circle of friends, nothing to do. But her brother had returned from Asia, and once again she was forced by her notions of familial duty to spend several long months looking after him when she really wanted to be on her way. She passed the time studying anthropology and fish. Finally Charles left England again, and she was able to set off for West Africa in December 1894.

This is the trip that made Mary Kingsley's reputation as a great traveler. She started in Nigeria, in the vast tangle of mangrove swamps where the Niger River splits up into many mouths and meets the sea. Here she had African guides paddle her in canoes up the streams and across the lagoons so that she could collect fish. She was impressed by what she thought must be the largest swamp region in the world, and she found in it a melancholy beauty. "In its immensity and gloom," she wrote, "it has a grandeur equal to the Himalayas."

Fish were not the only creatures she met in the mangroves. She later wrote:

> Now a crocodile drifting down in deep water, or lying
> asleep with its jaws open on a sand-bank in the sun,
> is a picturesque adornment to the landscape when you
> are on the deck of a steamer, and you can write home

about it and frighten your relations on your behalf; but when you are away amongst the swamps in a small dug-out canoe, and that crocodile and his relations are awake—a thing he makes a point of being at flood tide because of fish coming along—and when he has got his foot upon his native heath—that is to say, his tail within holding reach of his native mud—he is highly interesting and you may not be able to write home about him—and you get frightened on your own behalf. For crocodiles can, and often do, in such places, grab at people in small canoes.

On one never-to-be-forgotten canoe trip, a crocodile got both its "front paws" over the edge of her boat. She was not sure whether it wanted to climb in or to tip her out, but neither possibility appealed to her. She gave the crocodile "a clip over the snout" with her paddle, and it swam away.

Leaving Nigeria, Kingsley moved south to Libreville, a French colony in Gabon. Her goal was the Ogowé River south of Libreville. The Ogowé runs close to the equator and is the largest river between the Niger and the Congo rivers. There was a mission settlement 130 miles (208 kilometers) upriver and a trading post called Ndjole some distance further, but beyond Ndjole the Ogowé was blocked by rapids that the river steamers could not pass. Very little was known about the land and people along its upper course. Kingsley wanted to look for new species of fish there. She also hoped to pay a visit to an elusive tribe called the Fang (she called them Fans, because she was afraid that "Fang" would sound too ferocious to her English readers). The Fang were rumored to be cannibals; she wanted to investigate their culture and replace rumors with facts.

She set off upriver from Ndjole in a canoe, with several African guides and a pile of trade goods. It was a wild ride. The rapids were rough, and more than once she was tipped into the water and had to scramble up the steep, jungle-covered bank. But the river was lovely and she found many fascinating fish. After this excursion she returned partway down the river to the mission station, Lambaréné (later to become known around the world as the site of Dr. Albert Schweitzer's hospital). From there she set off northward into unexplored country to look for the Fang. She soon found them.

The first sign that the Fang were nearby was a pit that opened suddenly under Kingsley's feet. She went crashing down about 15 feet (5 meters) and landed on nine sharp wooden spikes, but she was not hurt; her bulky clothes had protected her. "It is at times like these you realise the blessings of a good thick skirt," she said. She refused to wear male clothing, as some women adventurers did; she always traveled in a black wool dress because, she felt, "you have no right to go about Africa in things you would be ashamed to be seen in at home." This attitude seems ridiculous today, but she was quite serious about it. For example, she knew that she would be more comfortable if she wore an African hat with a brim to shield her eyes and neck from the sun, but she felt it would not be proper, so she stuck with her English lady's cap and veil.

The pit was an animal trap, and just beyond it was a Fang village. As Kingsley and her guides approached the village, she experienced one of her rare moments of fear. The Fang men surrounded her party, muttering angrily and brandishing their weapons. Kingsley reported that "it was touch-and-go for twenty of the longest minutes I have ever lived, whether we fought—for our lives, I was going to say, but it would not have been even for that, merely for the price of them." But just when things looked worst, one of her guides recognized a Fang he had known some time earlier. This man greeted them, and the tension passed. Kingsley was granted the status of guest and given a house for the night.

That evening she was bothered by a terrible smell that seemed to be coming from somewhere in the house. She tracked it to a small bag that was hanging from the roof. Cautiously she emptied the

Kingsley sets out on the Ogowé River in search of "fish and fetish."

A Fang family. Kingsley called the Fang people Fans because she felt that their true name might strike her readers as rather bloodthirsty; the Fang were cannibals, after all.

contents of the bag into her hat, and when she saw what these contents were, she realized that the Fang were cannibals after all. "They were," she wrote, "a human hand, three big toes, four eyes, two ears, and other portions of the human frame. The hand was fresh, the others only so-so, and shrivelled." She decided not to take her usual solitary evening stroll in the forest that night.

Kingsley spent at least a week, maybe longer, in the Fang country, traveling north from village to village. The trip was tiring, not just because it involved walking for many miles through hilly, muddy rain forest but also because Kingsley was expected to act as doctor in every village they visited. Furthermore, she had to serve as a judge in the settlement of dozens of long-winded disputes between the Fang and her guides. Yet it was also extremely interesting.

Kingsley had the opportunity to study many aspects of Fang life—she even saw an elephant hunt and the feast that followed it. Using her guides as interpreters, she pursued her study of "fetish," questioning the Fang about their beliefs and customs. She was particularly interested in magic and witches; by this time she was something of an authority on West African witchcraft. She came to like the Fang very much. She could not approve of their cannibalism, but she admired their vigor and independence.

After crossing about 200 miles (320 kilometers) of forest and swamp, Kingsley and her group came to the Rembwé, a stream that flows into the Gabon River, which in turn flows down to Libreville at the coast. Here she said farewell to her Fang companions, who went back into the forest, and found a trader to take her downriver. She proudly reported that she had become such a good pilot that he let her steer the canoe at night.

Now that Kingsley had achieved such success with "fish and fetish" on the Ogowé and among the Fang, she was in a light-hearted mood. The Rembwé River trip was one of her happiest experiences:

> Indeed, much as I have enjoyed life in Africa, I do not
> think I ever enjoyed it to the full as I did on those nights
> dropping down the Rembwé. The great, black, wind-
> ing river with a pathway in its midst of frosted silver
> where the moonlight struck it; on each side the ink-
> black mangrove walls and above them the band of

star and moon-lit heavens....Ah me! Give me a West
African river and a canoe for sheer pleasure.

From Libreville Kingsley worked her way back up the coast. She
had been in Africa nearly a year. She was ready to go back to England
to turn in her fish specimens at the British Museum and write a book
about her trip. First, though, she had one more African adventure.
She stopped in Cameroon, which was then a German colony, to climb
a mountain called Mungo Mah Lobeh, or "Throne of Thunder" (now
called Mt. Cameroon).

Standing at the point where the coast of West Africa turns sharply
south, Mungo Mah Lobeh is the highest point on the western side
of Africa (13,353 feet or 4,673 meters). Kingsley wanted to see the
view. Her German hosts urged her to reconsider; some of them had
tried the climb, and they knew how dangerous and difficult it was.
But Kingsley could be quite stubborn in her quiet, humorous way,
and she climbed the mountain.

It took her four days. She ran out of food and water, got lost several
times, was drenched by rain and fog, and was deserted by her porters
before she reached the top—but she reached it. To her disappointment, though, she never saw the magnificent view she had expected.
Fog and clouds wrapped the peak the whole time she was on it. Still
she was glad to have made the climb. She was the first European
woman to do it, and only a few men had preceded her.

Kingsley reached London in November 1895. The fish experts at
the British Museum were happy with her specimens. She had
discovered half a dozen new varieties; one species, *Ctenopoma
kingsleyae,* was named after her. She began writing and produced
dozens of articles for magazines, as well as a narrative called *Travels
in West Africa,* which was published in 1897. It was tremendously
popular, and she followed it with *West African Studies* in 1899.

But even before her books appeared, Mary Kingsley was a
celebrity. Word of her exploits had reached England while she was
still in Africa. Upon her return she found herself in great demand as
a lecturer and a dinner-party guest.

Soon she became rather controversial. Most British people at the
time felt that Africans were fundamentally inferior to Europeans.
Whites often said that the Africans were like children—sometimes

Kingsley was proud of her
scientific studies of African fish.
These are some of the drawings
she made of her specimens;
she discovered many new
species, and several of them
were named for her.

black people were even compared with animals. The British
government's policy, supported by the missionary societies, was to
take these "primitive" people in hand and turn them into "decent,
law-abiding Christians."

Mary Kingsley was extremely patriotic. She had no quarrel with
the idea that African territory should become part of the British
Empire; in fact, she wanted the British to have more influence in Africa
than the French or the Germans. But she believed that the British
and the Africans should be equal partners, instead of the British
forcing their will on the Africans. She was especially critical of the
missionaries, who wanted to turn all Africans into Christians, for she
believed that the Africans' traditional beliefs were as worthy of respect
as anyone else's. She complained that Europeans seemed to regard
the African mind as a jug, which they could empty of its original
contents and fill up with their own "second-hand rubbishy white
culture."

These views were considered bizarre. Kingsley's knowledge and
experience were recognized, but her opinions had no effect on the
course of British colonial policy. Today, however, Kingsley is
regarded as having been a traveler ahead of her time—surprisingly
open-minded, free of prejudice, and free also of the moral superiority
that most nineteenth-century Europeans felt in foreign lands.

Kingsley had a busy public life after she returned from Africa, but
her private life was empty. She once said that it saddened her that
she had never known love. In the late 1890s she met a young military
officer named Mathew Nathan, who was also interested in West
Africa; he was to become governor of Sierra Leone. Kingsley became
very fond of Nathan, and may have been in love with him. She shared
her inmost feelings with him, telling him in a letter, "The fact is I am
no more a human being than a gust of wind is. I have never had
a human individual life. I have always been the doer of odd jobs—
and lived in the joys, sorrows, and worries of other people. It never
occurs to me that I have any right to do anything more than now
and then sit and warm myself at the fires of real human beings."

Whatever Kingsley's feelings for Nathan, he felt only friendship
for her. He went to Africa and her brief episode of emotional
excitement ended. Not long afterward, in March 1900, she set off on
her third and final trip to Africa. At this time South Africa was in the

(continued on page 89)

MARY KINGSLEY'S AFRICAN TROPHIES.

Mary Kingsley's friends begged her not to go to West Africa, a region they thought was dangerous beyond belief. One gentleman advised her to visit Scotland instead. And when she tried to learn some phrases in West African languages, she was startled to note that the first sentence in the phrasebook was "Help, I am drowning!" But Kingsley's heart was set on seeing Africa, and she would not be swayed. Like many other women, she discovered a whole new realm of independence and achievement in travel. The artifacts and biological specimens she brought home were more than scientific curiosities—they were symbols of her newfound freedom.

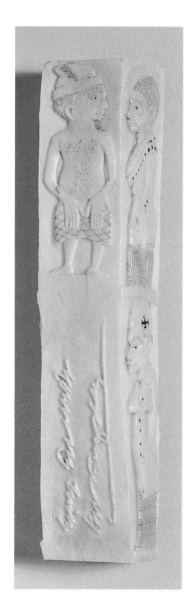

Ivory was greatly sought after by European traders in Kingsley's day. Most of it was carried out of Africa and fashioned into all sorts of things, from piano keys to gold-handled toothpicks. But the African artisans were master ivory carvers. This piece was carved for Kingsley—with human figures and her signature—somewhere along the Congo coast.

A mask of brass and copper from Nigeria; the beads are of coral, which was a valuable trade commodity in West Africa. Such masks were worn by tribal leaders, either around the neck or attached to a hip belt. Kingsley was deeply interested in fetishes, or objects that served as symbols of power or religious mystery.

A crocodile mask. To the Oba and Edo people of Nigeria, the crocodile represented power and ferocity. It was associated with Olokun, god of the waters. Kingsley had several spirited clashes with Olokun's reptilian representatives in the delta of the Niger River.

When she returned to England from her first West African trip, Kingsley brought with her a three-foot-tall statue of an idol that she called Muvungu. Studded with nails and ominously stained with dried blood, it stood in the front hall of her apartment, where it doubtless startled many a visitor.

In Africa Kingsley encountered people who made all their dishes, tools, and clothing from the materials at hand. Everyday items like this covered bowl were often carved to resemble animals or decorated with designs. But although factory-made cloth and skillets from England could not match the beauty of traditional African household items, mass-produced trade goods were beginning to transform African life.

"Hair-dressing is quite an art among the Igalwa and M'pongwe women," reported Kingsley. These peoples lived along the Ogowé River. Kingsley admired their meticulously plaited hairstyles, which probably looked much like modern cornrows.

A brass fan decorated with symbols and geometric shapes provided some relief from West Africa's oppressive heat. A less ornamented, but equally efficient, fan could always be improvised from a palm leaf.

The sansa, something like a small portable piano, had keys made of cane. Kingsley loved African music and tried her hand at every instrument she came across. One of her African acquaintances told her that she had "the makings of a genius for the tom-tom."

This bronze plaque from the Nigerian city of Benin depicts an African fish. Kingsley wrote, "I can honestly and truly say that there are only two things I am proud of—one is that Doctor Gunther has approved of my fishes, and the other is that I can paddle an Ogowé canoe." Gunther was the fish specialist who encouraged Kingsley to collect specimens for the British Museum.

(continued from page 80)

middle of the Boer War, which pitted the British against the Boers (Dutch colonists). Kingsley—always ready to answer the call of duty—felt that there must be some "odd jobs" for her to do there. She volunteered to work as a nurse in a hospital for Boer prisoners of war in Simonstown, South Africa.

She wrote to a friend that the hospital was "a rocky bit of the valley of the Shadow of Death." An epidemic of fever was raging, and the prisoners were dying in agony in their filthy beds. Kingsley did what she could to help, but soon she caught the fever. She died on June 3, 1900, and according to her wish was buried at sea off the African coast.

Mary Kingsley's career as an explorer was brief, crammed into the eight years she lived after her parents' deaths. But she managed to travel for two of those years, to study extensively, and to write several long books and many articles.

Kingsley always traveled with a purpose, with an "odd job" of some sort to do, but what she really wanted from travel was peace. Unlike Isabella Bird Bishop, who gloried in the opportunities for individuality and independence that travel offered her, Kingsley sought escape from an unfulfilling life, not just from a stifling environment. One of her happiest memories was of a night when she crept away alone to watch the stars along the Ogowé River. She wrote:

> In the darkness round me flitted thousands of fire-flies and out beyond this pool of utter night flew by unceasingly the white foam of the rapids; sound there was none save their thunder. The majesty and beauty of the scene fascinated me, and I stood leaning with my back against a rock pinnacle watching it. Do not imagine it gave rise, in what I am pleased to call my mind, to those complicated, poetical reflections natural beauty seems to bring out in other people's minds. It never works that way with me; I just lose all sense of human individuality, all memory of human life, with its grief and worry and doubt, and become part of the atmosphere. If I have a heaven, that will be mine.

CHAPTER 6

Alexandra David-Neel: To the Forbidden City

Tibet! As soon as Europeans began traveling in Asia, they heard rumors of this hidden kingdom. Sealed·off from the rest of the world on a high plateau between China and India, Tibet was guarded on the north by the Gobi Desert and the Takla Makan Desert, two of the bleakest and most forbidding places on earth. On the south its ramparts were the Himalayas, the "Abode of Snows," the highest mountains in the world. From very early times Tibet was a symbol of mystery. It was the farthest of faraway places—the end of the earth.

Tibet was a theocracy, a nation in which the church is also the government. Its religion was Buddhism, and it was ruled by a caste of high priests called lamas. The highest-ranking lama was the Dalai Lama, who was both the leader of the Tibetan Buddhist faith and the head of state. Tibet was said to be the center of all sorts of mystical learning, which made it even more interesting to curious Westerners. At various times in its history it was under Chinese or Mongolian rule. But at all times the rulers of Tibet, like those of other Himalayan kingdoms, kept their country off limits to Westerners.

Still, a few Europeans managed to get there. Marco Polo may have passed through part of Tibet during his Asian travels in the thirteenth

Wearing felt-soled Tibetan boots and a pilgrim's robe, with a 108-bead Buddhist rosary around her neck, David-Neel sits among nuns of the Red Hat order, one of several Buddhist sects in Tibet.

century. Starting in the seventeenth century, a few Christian mission-aries, enduring great hardship, managed to reach Tibet. For the most part, though, Tibet remained a mystery to the rest of the world. Lhasa, its capital city, was the greatest mystery of all. It was called the For-bidden City because foreigners were strictly forbidden to enter it.

By the late nineteenth century, the British were in control of India. They became increasingly curious about Tibet, and soon a handful of daring explorers began to probe its borders. Lhasa remained inaccessible for many years; not until 1904 did a British explorer reach the Forbidden City, and he succeeded only because he came at the head of an army.

Tibet attracted women explorers as well as men. In the early 1870s Nina Mazuchelli, the wife of a British official in northeastern India, persuaded her husband to take her on a wildly impractical expedition to Mt. Everest, which straddles the southern border of Tibet. They passed through unexplored territory in the Himalayan kingdom of Sikkim and came close to Tibet before they were forced to turn back by bitter weather and lack of food.

In 1890 Isabella Bird Bishop toured Ladakh, or Little Tibet, on the southwestern border of Tibet proper. Two years later the first Western woman entered Tibet. She was Annie Taylor, an English missionary who lived and worked in China. She spent years traveling along Tibet's northern and eastern frontiers, longing to preach Christianity among the Tibetans. In 1892 she sneaked into Tibet with a Tibetan man who had run away from Lhasa and converted to Christianity.

Taylor crossed most of northern Tibet. She hoped to become the first Western woman to reach Lhasa, but she was discovered when she was still some distance from the capital and was banished from the country.

The first Western woman to reach the Forbidden City did not do so until thirty-two years later. Her name was Alexandra David-Neel, and her determination to reach Lhasa was matched only by her good fortune in surviving the journey. For David-Neel, the trip to Lhasa was more than a geographical achievement—it was a spiritual adventure. She combined the restlessness of an explorer with the curiosity of a student and the yearning soul of a religious pilgrim. "Adventure," she once said, "is my only reason for living."

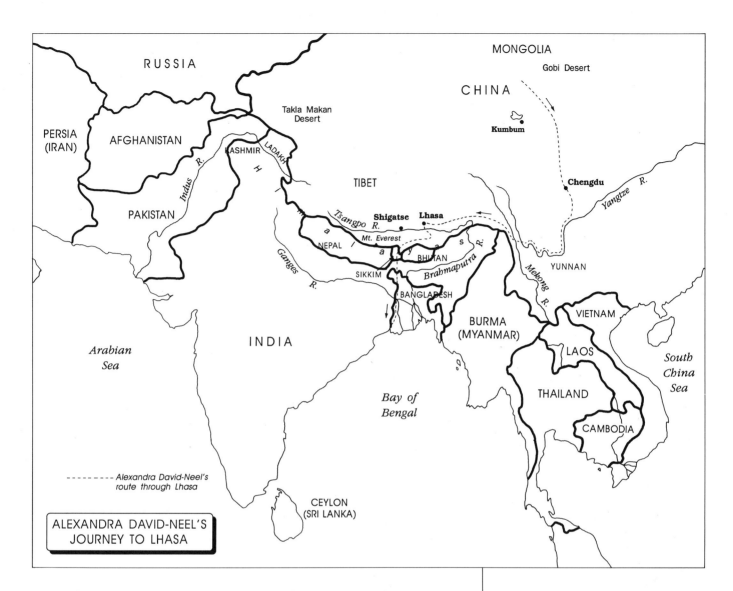

ALEXANDRA DAVID-NEEL'S
JOURNEY TO LHASA

- - - - - - - - Alexandra David-Neel's
route through Lhasa

She was born Alexandra David in Paris, France, in 1868. Her father was a schoolteacher and political activist. Her mother, twenty years younger than her father, had no intellectual or political interests, and Alexandra later recalled that their family life was cold and sterile. She was an only child who felt that her parents gave her little love, although she was well taken care of in every outward sense.

From her earliest childhood she had the urge to roam. She wrote:

> Ever since I was five years old, a tiny precocious child
> of Paris, I wished to move out of the narrow limits in
> which, like all children of my age, I was then kept.
> I craved to go beyond the garden gate, to follow the
> road that passed it by, and to set out for the unknown.

She was a solitary child who loved to read, especially books about travel and adventure. From them she gradually gained a mental picture of Asia: an ancient land of mountains and jungles and palaces, of wisdom and mystery. It seemed to call to her. Later in life, when she went to Asia, she found that this exaggerated image was a romantic fantasy. But she loved the reality that she found in its place.

At some point David discovered the Theosophical Society, an organization that taught certain religious and philosophical doctrines based on the religions of Asia, especially Buddhism. She began to make a serious study of Buddhism—a project that was to last for the rest of her long life. Her favorite place in Paris was the Musée Guimet, a museum of art and religious objects from the Far East. She spent many hours in its reading room, under the benign gaze of a huge statue of the Buddha, poring over books about Asia and Buddhism.

When she was twenty-three years old she was able to make a short trip to India with money from her godmother. She traveled across the country by train, expecting to see lush forests, holy men meditating under trees, and flowers and spices everywhere. What she found was heat, dust, crowds, and poverty mixed with splendor. It was confusing and exhilarating. She stayed until her money ran out.

It was now time for her to begin earning her own living, and she turned to the operatic stage. She had a lovely voice and had studied music as a child. Now she began singing again, working as hard as she could—the way she did everything. Soon she was giving well-regarded performances in popular operas. In 1896 she went to the French colony of Indochina (present-day Vietnam) as the leading soprano for a touring opera company. In the years that followed she also sang in Greece, North Africa, and the provinces of France. She might have made a lifelong career as a singer, but she damaged her voice by singing too much, so she switched to writing. She produced several novels, some political tracts, and many magazine and newspaper articles on Asian religion and culture.

She never expected to marry. She believed that a woman surrendered too much of her freedom when she tied herself to a man, and she did not want to have children. Yet she sometimes longed for companionship and for a feeling of security. At the age of thirty-six she married Philippe Neel, a French engineer who worked in Tunis, North Africa.

Theirs was a curious marriage—she once called it a "heart-rending comedy." They felt a strong attraction to each other and regarded each other as friends, but they had different ideas about marriage. Alexandra wanted a comrade who would support her emotionally and financially but would respect her desire to lead her own life; Philippe wanted a more conventional marriage, with children.

The first several years were stormy. Alexandra spent most of her time in Paris, and their marriage consisted mainly of agitated letters back and forth. Over the next few years they experimented with living together and apart, and Philippe came to understand that Alexandra was simply not cut out for domestic life. Alexandra entered a prolonged period of depression and confusion, caused partly by her marriage and partly by the death of her father. Finally she realized that what she really wanted to do was return to India to advance her study of Buddhism. She left in 1911 with Philippe's blessing—and with money he provided for her. He did not know that she would be gone for fourteen years.

In the years that followed, Alexandra David-Neel, as she was now known, became a personage of some importance in Buddhist circles in Asia. She traveled, lived, and studied in the Himalayan kingdoms of Nepal, Bhutan, and Sikkim (today Sikkim is part of India). These countries border on Tibet, and she loved to gaze across the snowy peaks to the forbidden plateau beyond. Many Tibetans lived in the border kingdoms, and from them she learned the Tibetan language.

She met many high-ranking Buddhists, who were impressed with her knowledge and spiritual dedication. They permitted her to wear the ceremonial robes of a lama. In 1912 she became the first foreign woman to meet a Dalai Lama, the spiritual leader of all Tibetan Buddhists. The Dalai Lama fled to India during a brief Chinese invasion of Tibet, and while he was there he spoke with David-Neel and encouraged her to keep studying Tibetan. She took this as an omen that someday she would enter Tibet.

Around this time the lamas at a Sikkimese monastery assigned a fifteen-year-old monk to be David-Neel's servant. His name was Yongden, and he spent the rest of his life with her, helping her with her travels and studies. He was a loyal companion; she adopted him as her son and heir.

All this time Philippe had been writing to her, urging her to come home. She kept putting him off and then, in 1914, World War I broke out. It was no longer safe to travel back to Europe. David-Neel did

A Buddhist tangka, or sacred painting, used as a guide to meditation. The white-capped peaks represent the Tibetan Himalayas, but they also symbolize the obstacles a person must overcome to achieve enlightenment. David-Neel collected many tangkas and other Buddhist artifacts.

just the opposite—she ventured across the border on a secret trip into a remote part of Tibet to visit a community of Buddhist nuns.

The nuns told her about a very learned, holy man who lived in a cave in Sikkim, and she became his student for a year. On his orders she spent the winter of 1914-15 in a Sikkimese cave at 13,000 feet (3,965 meters) of altitude, fasting, praying, and reading. Then she went back to Tibet. This time she visited the city of Shigatze, where there is an important Buddhist monastery. The Tibetans she met did not seem to mind her presence very much; in fact, she was entertained as an honored guest. Upon her return to Sikkim, however, the British colonial government ordered her to leave India. The British jealously policed the Himalayan frontier and were angry that she had crossed the border without their permission.

David-Neel was furious with the British. She became determined to defy their orders and somehow get to Lhasa. The only possibility was to enter Tibet from the north, from China, as Annie Taylor had done. After leaving India, David-Neel made a tour of Buddhist monasteries in Burma (today called Myanmar), Japan, and western China. In 1918 she and Yongden arrived at an ancient Buddhist monastery called Kumbum, just north of Tibet. They spent more than two years there while David-Neel translated old scrolls and documents. But because she was a woman she could not join the holy order, and eventually they had to move on.

David-Neel and Yongden spent the next several years wandering around western China, in the Gobi Desert and along the Tibetan border. It was a trying time for David-Neel. The letters she wrote to Philippe show that she was desperately poor. China was torn by civil wars; soldiers as well as bandits prowled the frontier regions, and she was often in great danger. Once a mule driver tried to rape her, but she beat him off with a whip and had no further trouble of that sort. Occasionally she felt tired and discouraged, and at times she dreamed of returning to a comfortable life in France, but the dream of reaching Lhasa was stronger, and she never gave up.

She and Yongden tried in 1921-22 to enter eastern Tibet by way of the Chinese city of Chengdu. At this time both the Chinese and the Tibetans wanted to keep foreigners out of Tibet, and the travelers were turned back at the border. Wrote David-Neel, "I took a silent oath to renew my attempt, if necessary, ten times." In the end, only one more attempt was needed—but the journey was a long one.

In 1923 they started from the Gobi Desert, north of Kumbum, and traveled south once more to Chengdu. Then they continued south and west in a long curve that took them across the Yangtze River and deep into Yunnan, China's southernmost province, along the border with Burma. From there they followed the Mekong River up and up through the eastern Himalayas and onto the high, cold Tibetan plateau. They had avoided the main caravan route from China to Lhasa and had instead taken a very obscure and difficult route, hoping to avoid detection. They had to cross several mountain passes that were more than 18,000 feet (5,490 meters) high.

Before entering Tibet they got rid of their baggage and disguised themselves as Tibetan beggars. David-Neel dyed her hair black with ink and darkened her skin with charcoal. Yongden posed as her son, a young lama. In order not to arouse suspicion, they carried no blankets or extra clothes, only a small tent and a teakettle. David-Neel hid a money bag and a revolver in her clothes. Each of them had a begging bowl and a pair of chopsticks. They ate what the poorest Tibetans ate—boiled barley and tea with a little butter and salt in it. That was on a good day; on a bad day they went hungry.

They covered the distance from China to Lhasa on foot, and the journey was a slow one. At least 500 to 600 miles (800 to 960 kilometers) of it was through regions that no Westerner had ever seen. David-Neel was able to make many important observations about the people and geography of eastern Tibet. Among other things, she discovered the source of the Po River, a tributary of India's mighty Brahmaputra. She also gossiped with old women in the villages and learned the local legends and customs. But in spite of these diversions the trip was remarkably difficult. More than once she and Yongden came close to starving or freezing to death, and both of them got sick with some sort of fever.

On January 1, 1924, Alexandra David-Neel saw in the distance the gleaming golden roofs and towering red and white walls of the Potala Palace, the Dalai Lama's home, which stands on a high hilltop in Lhasa. With dust swirling around her and tears stinging her eyes, she passed beneath an elaborately carved gate and entered the Forbidden City.

The victory had cost her dearly. All her life she had been plump, with a smooth, round face. But a photograph taken of her sitting before the Potala Palace in Lhasa shows a gaunt, withered face that

This controversial and mysterious photograph shows an emaciated David-Neel (center) sitting with Yongden and a Tibetan girl in front of the Potala Palace in Lhasa in 1921. Some critics have questioned its authenticity, pointing out that David-Neel did not have a camera. Yet there was at least one photographer in Lhasa who could have taken the photo. If it is genuine, it shows how profoundly David-Neel was weakened by her harsh and dangerous journey.

looks like a skull and a body as frail as a handful of sticks. This photograph has always been something of a mystery. One critic who later claimed that David-Neel lied about her journey to Lhasa said that the photo was a fake. David-Neel could not have taken it, for she did not take a camera with her. But she included the photo in her 1927 book about the trip, and she appears to have had it in her possession when she left Lhasa. The most likely explanation is that it was taken by a Nepalese photographer who is known to have lived and worked in Tibet in the 1920s. Pilgrims to the holy city may have bought souvenir photographs of themselves in front of the Potala Palace. Whatever the photograph's origin, the woman in it is definitely Alexandra David-Neel, looking tragically haggard and worn. In her published account of the trip she downplayed some of the difficulties she had experienced, making the journey sound more lighthearted than it really was. Privately she admitted to Philippe that she had come very close to death. She was fifty-five years old.

She and Yongden remained in Lhasa for two months. Still posing as a beggar-woman from a remote province of Tibet who had come to Lhasa on a pilgrimage, she managed to tour the Potala Palace by trailing behind a group of true pilgrims. The Potala was built in the seventeenth century and contains hundreds of chambers, many of them filled with treasures and sacred relics. David-Neel was the first Westerner to see some of these rooms.

It was dangerous to stay in Lhasa too long. Besides, David-Neel was now eager to tell her story to the world. She and Yongden slipped quietly out of the capital and headed south along the caravan trails that led through the Himalayas to India. She intended to show the British authorities there that they could not keep *her* out of Tibet.

She caused a sensation when she arrived in India in August 1924. Her journey was acclaimed as an extraordinary feat. As soon as she had recovered a little strength, she sailed for home after an absence of fourteen years. By May 1925 she and Yongden were settled in southern France with the hundreds of Tibetan books and manuscripts that she had shipped to Philippe over the years.

My Journey to Lhasa was published in *Asia* magazine in 1926 and as a book in 1927. It immediately became one of the most popular and successful travel books ever written. David-Neel was recognized as the world's leading authority on Tibet. She always pointed out, however, that her knowledge came not from a single year's travel but from many years of study.

In the decades that followed, David-Neel wrote more than twenty books about Tibet, Buddhism, and her experiences. She and Yongden lived near Digne, France, in a house that she called Samten Dzong (Tibetan for Fortress of Meditation). Philippe remained her friend and helper until his death in 1941, but he did not live with her.

David-Neel traveled to Asia again in 1936, at the age of sixty-eight. This time she remained for only eight years. During World War II she lived in southern China near the Tibetan frontier; she had to leave in 1944 when the Japanese were on the verge of invading the region. She went back to Samten Dzong and there she stayed. Her faithful friend Yongden died in 1955. Alexandra David-Neel lived on—studying, writing, and introducing Westerners to Buddhism and Tibetan mysticism—until her death in 1969, a few weeks short of her 101st birthday.

CHAPTER 7

Marguerite Baker Harrison: Born for Trouble

When Marguerite Baker Harrison published the story of her life, she titled the American edition *There's Always Tomorrow.* But in England it was called *Born for Trouble,* and Harrison's troubles were spectacular indeed. But so were her triumphs. In a single decade, from 1915 to 1925, she achieved distinction as a reporter, spy, traveler, and pioneering film producer.

Marguerite Baker was born in Baltimore, Maryland, in 1879. Her father owned a shipping company, and as a child and young woman she made many trips to Europe. She learned to speak French and German, skills that would later be useful when she took up espionage. Harrison attended an exclusive girls' high school near Baltimore; she also spent much time in her grandfather's library, reading his collection of books about history, exploration, and travel.

If the young Marguerite Baker dreamed of travel and adventure, those dreams receded into the background when she fell deeply in love with a Baltimore man named Thomas Bullitt Harrison. She married him in 1901 and entered quite happily into domestic life. They had one son, a boy whom they named for his father.

In Syria and Iraq, Marguerite Baker Harrison traveled by camel, the "ship of the desert."

Harrison, a housewife-turned-spy, combined exploration with moviemaking in the 1920s.

Tragedy struck in 1915, when Thomas Harrison died after a serious illness. His death left the family not only penniless but in debt. Marguerite Harrison had no professional training or experience, but she believed that she could be a reporter. She talked the Baltimore *Sun* into giving her a job. Harrison started out as the assistant editor of the society page, writing articles about tea parties and fashion shows. On her first day on the job, she did not even know how to use her typewriter. But her writing was so good that the paper soon made her its music and drama critic, and then gave her her own weekly column.

By 1918, however, Harrison was impatient with this work. World War I was drawing to a close in Europe, and she wanted to see what was going on over there. In particular she was curious about Germany, which she had visited and loved years before. But it was wartime, and Germany was a battlefield. American civilians could not just pick up and travel to an enemy country. Harrison decided to get the government to send her there. She applied to the Military Information Division (MID) of the U.S. War Department for a job as a spy. On her application she wrote, "I have absolute command of French and German, am very fluent and have a good accent in Italian and speak a little Spanish. Without any trouble I could pass as a French woman, and after a little practice as German-Swiss."

The MID investigated her. One of its reports said, "She remarked she is fearless, fond of adventure, and has an intense desire to do something for her country." She was accepted as a secret agent. The war ended before she reached Europe, but the MID told her to poke around Germany, under the cover of newspaper reporting, and to investigate new political and economic developments. She put her son in boarding school and began her espionage career.

In 1920 the MID sent her to Russia. This was a dangerous assignment. The communist revolutionaries who had taken over the country a few years earlier were just beginning to form the new Soviet Union, and they did not want any interference from the rest of the world. A wall of silence and secrecy was being built around the Soviet state.

In Poland on her way to Moscow, Harrison met an American pilot named Merian C. Cooper. They danced and chatted at a Red Cross club and then went their separate ways. Harrison could not have imagined the adventure they would share a few years later.

Harrison reached Moscow safely, but after eight months there she was arrested by the Soviet secret police. They knew that she had been smuggling information about Soviet affairs out to the MID, and they threw her into Lubianka Prison, where political prisoners and spies were held. The prison building had once been the headquarters of an insurance company, and as she was hustled through the doorway she looked up and saw an ominous motto on the wall: "It is prudent to insure your life."

She was the first American woman to be held in the Soviet prison system, whose horrors would be described decades later by the Soviet writer Alexander Solzhenitsyn in his book *The Gulag Archipelago*. Harrison's book *Marooned in Moscow* tells of her own experiences as Prisoner Number 2961. Conditions were dreadful and her health deteriorated. The worst thing was the uncertainty of her position. She thought she would probably spend the rest of her life in prison.

Fortunately, the U.S. government was able to arrange her release after ten months. She returned to the United States and her son, and she took up lecturing and writing. In 1922 she left on a trip to Asia to gather material about the social and political upheavals that were occurring in Japan and China.

She crossed the Pacific Ocean by steamship, landing in Japan. Then she went on to Korea and China. She had planned to return the same way, but she was seized with the desire to cross Mongolia and Siberia. As she was no longer a spy, she believed she could enter the Soviet Union safely and make her way across that country to Europe, returning home by way of the Atlantic Ocean.

The trip across Mongolia's Gobi Desert was an adventure. She traveled in the car of a British fur trader, but all around her were caravans of swaying camels and rifle-toting nomads. Urga, the old capital of Mongolia, had just been captured by a Soviet army (the city's name was later changed to Ulan Bator). She found it a busy place, a crossroads of trade and travel, full of Mongol princes in furs, Soviet soldiers in drab uniforms, yellow- and scarlet-clad Buddhist monks from all over Central Asia, and Chinese merchants in long robes. Few Western women had visited Urga.

Harrison proceeded westward into the Soviet Union. The Soviet consul in Beijing, China, had given her a visa, so she felt safe. But not long after entering Siberia she was arrested, charged with

Merian C. Cooper was one of Harrison's partners in the Persian expedition. He and Ernest Schoedsack, the third of the three partners, later went on to make one of the best-known movies of all time, *King Kong*.

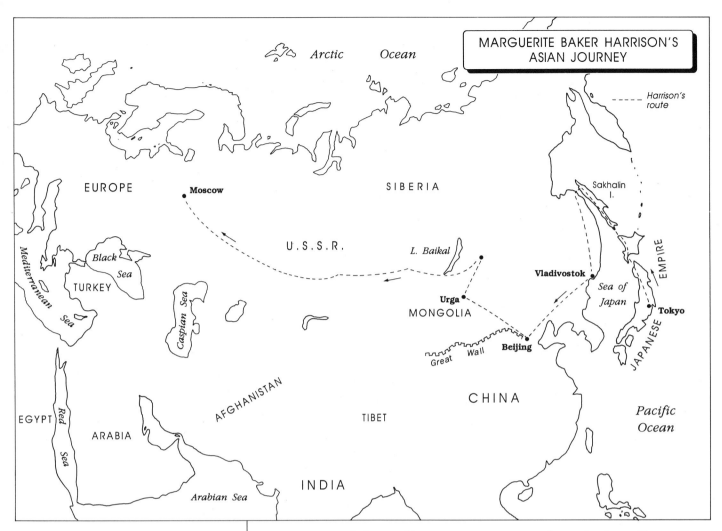

Harrison's
route

Arctic   Ocean

EUROPE   Moscow   SIBERIA   Sakhalin I.

Black
Sea
TURKEY   U.S.S.R.   L. Baikal   Vladivostok   Sea of Japan   EMPIRE

Mediterranean
Sea   Caspian Sea   Urga   Tokyo
          MONGOLIA   JAPANESE

EGYPT   AFGHANISTAN   Great   Wall   Beijing

ARABIA   TIBET   CHINA   Pacific Ocean

Red Sea   INDIA

Arabian Sea

espionage, and shipped by train to Moscow, where she found herself once again in the hands of the secret police. She was terrified—and also angry. Her first arrest had been bad enough, but she admitted that she had broken the Soviet law by spying. This time she actually had done nothing wrong. She was afraid, however, that she would not be able to make her captors believe her.

She was held in prison for ten weeks while the secret police tried to pressure her into spying for them in the United States. She refused. Her release was a matter of pure good luck. An American official who happened to be in Moscow caught a glimpse of her and recognized her. The U.S. government demanded that she be set free, and the Soviets, who needed food and other aid from the United States, were forced to comply.

Harrison reached home in March 1923. But although she settled down in New York City to write a book, she had not yet had her fill of adventure.

Soon afterward she ran into Merian Cooper, the pilot she had met in Poland. He had just made a film in Africa with a cameraman named Ernest Schoedsack, and he was full of enthusiasm for moviemaking.

The film industry was growing fast, and Cooper felt that travel films might be very successful. Harrison agreed, and decided to join with Cooper and Schoedsack to make a film about some exotic part of the world. They settled on an original and challenging subject: the life of a nomadic tribe far from Western civilization. Harrison raised $10,000 for the project. She was to be the movie's financial backer as well as one of its producers.

The next task was to find a suitably untouched and picturesque tribe of nomads. Harrison and her partners spent months on a difficult and frustrating journey through Turkey, Syria, and Iraq in search of the perfect tribe. The trip was a nightmare of bureaucratic red tape, delayed travel permits, medical crises, broken-down transportation, snowstorms, and disappointing nomads (Harrison called one group "a squalid, moth-eaten lot").

Finally the moviemakers agreed on the Bakhtiari, a nomadic people who lived in western Persia. Austen Henry Layard, a nineteenth-century explorer, had described the Bakhtiari as "arrant robbers and freebooters, living upon the plunder of their neighbors and of caravans." Jacob Bronowski, a twentieth-century British scholar, said of the Bakhtiari, "They are as near as any surviving, vanishing people can be to the nomad ways of ten thousand years ago."

There were about 30,000 Bakhtiari in Persia. The central drama of their lives was the seasonal migrations they made with their huge herds of livestock. In the winter they lived in the lowland plains along the Persian Gulf. In the summer, though, those plains become too dry and hot to support their sheep, goats, cattle, and horses. So every spring the Bakhtiari packed up all their possessions and drove their animals before them in a vast, slow-moving herd for more than 200 miles (320 kilometers) to their summer grazing grounds in the highlands of central Persia. Along the way they crossed six rugged mountain ranges and the Karun River, which was often flooded. They repeated the journey in the opposite direction in the fall.

Now that Harrison and the other filmmakers had decided on their subject, they had to win the approval of the Bakhtiari themselves. They learned that the Bakhtiari princes were camped near the town of Shushtar; the spring migration was about to begin. They raced to Shushtar and explained through an interpreter to the Il-Khani, the head of all the Bakhtiari clans, that they wanted to come along on the migration and make a record of it. They would eat, sleep, and travel just like the Bakhtiari. They had learned that the Bakhtiari clans followed five separate routes through the mountains, and they wanted to take the most difficult one, because it would make the most interesting record.

The interpreter laughed "until the tears came," said Harrison, but the Il-Khani granted their request. The filmmakers were told that they could travel with the Ahmedi clan. They were also told that no foreigner had ever followed the particular migration route they had selected. Thus they would be not just movie producers but explorers.

During her time in Persia, Harrison experienced a peculiar feeling of familiarity. She had felt it also in Russia and Turkey, but not in China or Japan. She wrote that when the Il-Khani arranged a river excursion, "It seemed the most natural thing in the world that I should be riding on a barge of goatskins down a mountain river. Somewhere, sometime, I knew that I had done it before." She decided that she must have been a Persian in an earlier lifetime.

In Shushtar the partners obtained mules and donkeys to ride and to use as pack animals. They were introduced to Haidar Khan, the head of the Ahmedi clan, who made it clear that he disapproved of the whole idea. But the Il-Khani's word was law, and Haidar Khan looked after them responsibly.

The migration began on April 16, 1924. Before dawn the 5,000 members of the clan rolled up their tents and hid them under piles

Harrison rode to Mongolia in a car, but modern transportation had its drawbacks. The car mired down in a swamp on the Mongolian steppe.

The American filmmakers join the Ahmedi clan of the Bakhtiari tribespeople for their migration across the Zagros Mountains to their summer pastures in the Persian highlands. No foreigner had ever followed this migration.

of stones to be retrieved in the fall. Kettles, cooking pots, rugs, and small children were strapped to the sides of horses. Some of the women and children rode, but most people walked, and most of them were barefoot. Slowly, with great clouds of dust and much noise, the herds were prodded into motion. The sheepdogs barked incessantly. It was, said Harrison, a "gorgeous colorful torrent of humanity."

The torrent flowed onward for weeks. When they came to the mountains, the tribespeople walked barefoot through snow and ice for days at a time. When they came to the fast-flowing, icy, treacherous Karun River, it took six days to get everyone across. The women and children rode on crude goatskin rafts; the men swam, using inflated goatskins as life preservers. They had to force the bawling, reluctant livestock every step of the way. Dozens of animals drowned, but no people were lost—not that year.

The worst part of the trip was the Zardeh Kuh, a mountain wall that rose steeply to a height of 15,000 feet (4,575 meters) at the end of the trail. Cooper described it as "black and yellow rock and snow and more snow, rising straight up until the peaks are lost in the clouds." The men chopped holes in the ice to serve as a precarious, zigzag ladder; it took four days just to prepare the trail. Then, slowly, the whole tribe inched up the mountain. Once they reached the summit, they ran and slid down the long slippery slope on the other side. Beyond it lay the green, flowery summer pastures. The trip had lasted forty-eight days. Harrison, Cooper, and Schoedsack were the first foreigners who had ever made the journey.

The filmmakers were met by the Bakhtiari princes, who were pleased and surprised that they had survived the ordeal. They were taken by car to Tehran, the capital of Persia, and from there they returned to the United States.

During the filming, Harrison and Cooper had disagreed about the approach they should take. Cooper felt that the movie should focus on the dramatic, magnificent spectacle of thousands of people struggling against the panoramic backdrop of nature. Harrison wanted more attention paid to the intimate details of the Bakhtiaris' lives—how they cooked, what they wore, how they interacted with each other around their evening campfires. In general, though, she felt that they had made a good and important movie. But when the movie studio released the film under the title *Grass* in March 1925, she was disgusted.

The film was silent (talking movies were not yet being made), but the studio had added subtitles that she felt were ridiculously out of harmony with the movie. For example, a Bakhtiari nomad is shown saying, "B-r-r, this water's cold!" when he steps into the Karun River. Harrison rightly felt that the beauty and dignity of the movie had been cheapened. She saw it only once.

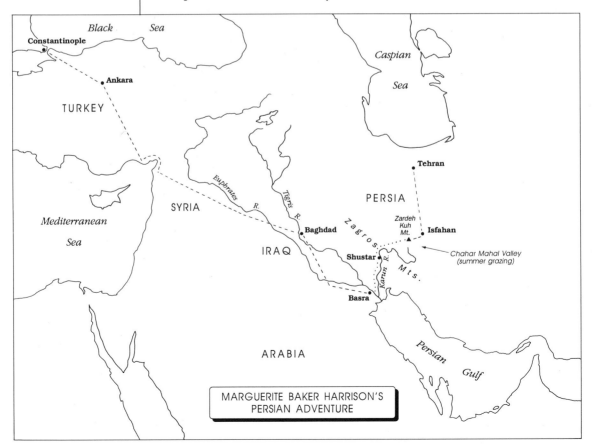

MARGUERITE BAKER HARRISON'S
PERSIAN ADVENTURE

The truth was that movie reviewers and audiences did not know what to make of *Grass*. Some people wondered why it did not have a big star like Rudolph Valentino in a leading role. Others thought it should have been a love story. People were simply not used to documentary films—they were a new phenomenon. But geographers and scientists loved *Grass*. They realized that film offered a revolutionary way to study, record, and teach others about distant lands and places.

Grass was a true innovation, a pioneering use of a new technology. It was the ancestor of educational television as well as of documentary films. Today it is recognized both as a cultural and geographic epic and as a landmark in film history.

Merian Cooper and Ernest Schoedsack went on to make more films together, including the classic monster movie *King Kong*. Harrison never made another movie, although she did live in Hollywood for a time with her second husband, an English actor named Arthur Middleton Blake, whom she married in 1926. Blake died in 1949, and Harrison returned to Baltimore to live near her son and his family. Late in life she made several trips to South America. She also visited Australia and traveled through much of Africa. She died in 1967 at the age of eighty-eight.

Harrison rode with the tribe during the migration. The Bakhtiari expected her to treat their illnesses and injuries, and she later wrote, "My experiences as a doctor did much to disillusion me about any ideas I may have had as to the romance and glamour of tribal life."

Not only was Marguerite Baker Harrison a pioneering filmmaker and an adventurous traveler, but she performed a service for all other women travelers. Frustrated at the way that women explorers were often treated as insignificant amateurs by the press and by the scientific establishment, in 1925 she joined with three female colleagues—Blair Niles, Gertrude Emerson, and Gertrude Mathews Selby—to form the Society of Women Geographers. With headquarters in Washington, D.C., the society offered women scientists, explorers, serious travelers, and travel writers a professional organization of their own. For the rest of her life, Harrison was prouder of this achievement than she was of *Grass*.

CHAPTER 8

Louise Arner Boyd: A Socialite in the Arctic

An elegant mansion on a beautiful estate in California, and a lonely wooden ship in the icy waters off a desolate Arctic shore—Louise Arner Boyd divided her life between these two wildly different worlds. Blessed with the money and freedom to do anything she wanted, she labored for years to contribute to the exploration of the region around the North Pole. Her greatest desire was to be taken seriously as an explorer, to be respected by the community of scientists and explorers whom she admired. She devoted her substantial fortune and her formidable energy to achieving this goal.

Boyd was born on September 16, 1887, at Maple Lawn, her family's home in San Rafael, California, just north of San Francisco. The house stood amid spacious lawns and gardens, for her family was wealthy. Her mother had inherited a fortune that was born in the California gold rush of the 1850s, and her father owned an investment company.

Louise Arner Boyd was the couple's third child; she had two older brothers. The Boyd family was part of San Francisco's high society at the end of the nineteenth century—an exclusive group of wealthy

Louise Arner Boyd with one of the eleven polar bears she shot on her second Arctic expedition. Soon her interests were to change from sportsmanship to exploration.

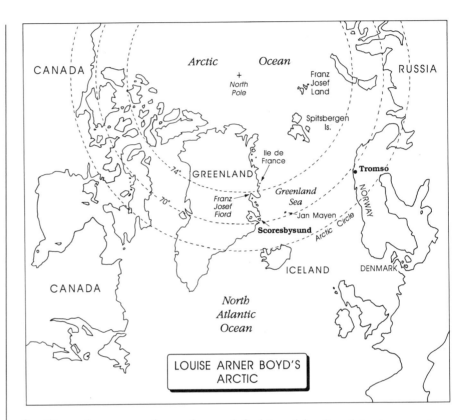

LOUISE ARNER BOYD'S
ARCTIC

families whose members dressed fashionably, lived in impressive homes tended by servants, and played a leading role in the city's business and political life. As a child Louise Boyd had a governess who gave her lessons at home. As a teenager she attended two private girls' academies—"finishing schools," as they were called—where young ladies were sent to complete their educations. She did not attend college.

Boyd was healthy and energetic, and she enjoyed outdoor life. She learned to ride and shoot at her family's ranch in the country. But she was not a thoroughgoing tomboy; she was proud of her good looks, loved stylish clothing, and enjoyed a sophisticated social life. Later, when she became an explorer, she tramped over the ice and tundra in men's trousers and hobnailed boots and slept in a tiny tent. Yet she took her maid with her on every voyage of exploration, and she told a reporter, "I may have worn breeches and boots and even slept in them at times, but I have no use for masculine women. At sea, I didn't bother with my hands, except to keep them from being frozen. But I powdered my nose before going on deck, no matter how rough the sea was. There is no reason why a woman can't rough it and still remain feminine." All her life she was proud of being able to endure cold, discomfort, and danger as well as any man could—when she had to. But she saw no reason to do so unnecessarily.

Boyd's childhood was privileged, but it was also rather lonely at times. The rest of her family suffered so many illnesses that her home life was always quiet; there was a nurse in the house at all times. Her older brothers had never been strong, and both died in their teens of rheumatic fever, a disease from which many young people perished in the nineteenth century. Her parents' health was also poor. Perhaps because of her family's perpetual ill health, Boyd developed an interest in nursing. She was to turn this interest to practical use as a volunteer nurse's aide in 1918, when an epidemic of influenza swept the country, and throughout her life she supported the work of the Red Cross.

In the subdued, sickroom atmosphere of her home, Boyd found companionship in books—especially books that widened her horizons. Years later she recalled that she had been "fond of geography from earliest childhood." Her favorite geography books were tales of travel and adventure in "high northern latitudes" close to the North Pole.

These were exciting years in which to become interested in the Arctic. Starting in 1897, when Boyd was ten years old, explorers from various nations made a series of attempts to reach the North Pole, first by balloon, then by sea or by foot across the northern ice. Each year, it seemed, a new expedition set off amid a welter of publicity, and the world waited anxiously to hear the results. Finally, in 1909, two men—Frederick Cook and Robert Peary—claimed to have reached the pole in separate expeditions. But although the race for the pole appeared to be over, much of the world north of the Arctic Circle remained unexplored, a mysterious white emptiness on the map.

Boyd's own experience of travel began in 1910, when she was twenty-three years old. Her parents took her on the Grand Tour that had become traditional for wealthy Americans, a year-long trip through Europe and Egypt. In the following years her father introduced her to the world of business, training her to manage his company. Her mother died in 1919, her father in 1920. She inherited the Maple Lawn estate and the Boyd Investment Company, and she found herself, at the age of thirty-two, free to choose her own road. Her choice was to travel.

She and another prominent socialite, Sadie Pratt, toured France and Belgium in 1920, noting the devastation that World War I had brought to Europe. They returned to Europe the following year and visited other countries, as many wealthy American tourists were doing. But for her next trip, in 1924, Boyd chose a more daring and unusual destination. At last she would see the far northern world she had been reading about for so long. She went to Norway and took a small tourist boat to Spitsbergen, a cluster of islands that lie in rugged isolation in the Arctic Ocean between Europe and Greenland, far above the Arctic Circle at the southern edge of the polar ice sheet.

Spitsbergen was a revelation to Boyd. The grandeur of the Arctic, and above all her first glimpse of the polar ice cap, thrilled her. She later wrote: "Far north, hidden behind grim barriers of pack ice, are lands that hold one spellbound. Gigantic imaginary gates, with hinges set in the horizon, seem to guard those lands. Slowly the gates swing open, and one enters another world where men are insignificant amid the awesome immensity of lonely mountains, fiords, and glaciers." Boyd had discovered in the north not only a new world but also a new purpose in life. As she gazed over the ship's rail at the glittering ice sheet that stretched away to the north, she said, "Someday I want to be way in there looking out instead of looking in." The cruise to Spitsbergen, she later said, "laid the foundation" for a career in exploration that would include seven Arctic expeditions.

The following summer she hired the *Hobby,* a Norwegian seal-hunting ship, and invited some friends to accompany her to Franz Joseph Land, a cluster of islands north of Russia, even farther into the Arctic than Spitsbergen. Being aboard the *Hobby* was particularly exciting for Boyd because the vessel had been the flagship of an expedition led by the Norwegian polar explorer Roald Amundsen, a hero of Arctic and Antarctic exploration. When Boyd took charge of the *Hobby,* it had just returned from carrying supplies to Spitsbergen for Amundsen.

The excursion to Franz Joseph Land was recreation, not exploration. The main pastime of the group was hunting—Boyd herself was said to have shot eleven polar bears and three seals. But she

also found time to experiment with a new hobby, photography. She shot a great deal of movie film and took 700 still photographs of the landscape. (Years later, during World War II, when the United States was sending supply ships to its Soviet allies through the Arctic waters above Norway, Boyd's photographs of Franz Joseph Land were a great help to the U.S. Navy, as they were the most complete record in existence of the islands' coastlines.) As far as we know, Boyd was the first woman to visit Franz Joseph Land. Upon her return home she told the San Francisco *Chronicle,* "I have got the Arctic lure and will certainly go north again."

Two years later she did so, chartering the *Hobby* for the summer season of 1928. She was to depart for Franz Joseph Land from Tromso, Norway. But she arrived in Tromso just in time to hear news that shook the world: Roald Amundsen, who had set forth toward the

Boyd became an expert photographer, skilled in the use of new equipment and techniques. Her pictures were so good that highly accurate maps could be made from them; these were the first maps ever made of parts of Greenland.

North Pole in an airplane to help rescue a stranded Italian expedition, was now missing, too. The government of Norway declared that no effort would be spared to find and save Amundsen. A search party of seasoned polar explorers, as well as news reporters from many countries, rushed to Tromso to begin the rescue operation. French, Danish, Italian, and Soviet vessels offered to help.

Boyd promptly offered her ship, crew, and supplies to help in the search. She asked only that she be allowed to take part in person. The Norwegian government gratefully agreed and assigned six army officers to the *Hobby,* which sailed from Tromso on July 1.

The *Hobby* went north to Spitsbergen, west into the Greenland Sea, and then east to Franz Joseph Land—a distance of more than 10,000 miles (16,000 kilometers) along the edge of the Arctic ice pack. The search was frustratingly difficult, because in the Arctic the eyes and mind can be tricked by deceptive effects of light on the ice. "Four of us stood watch around the clock," Boyd later recalled. "We would just stand there and look. Ice does such eerie things. There are illusions like mirages, and there were times we clearly could see tents. Then we'd lower boats and go off to investigate. But it always turned out the same—strange formations of ice, nothing more."

After more than three months, the search was called off. Amundsen was never found, although some of the stranded Italians were rescued from the polar ice. But in spite of the tragic nature of the search, Boyd enjoyed herself immensely. In her earlier voyage in the *Hobby,* her life had touched the edge of the world of serious polar exploration; now she had truly entered that world. During the search she met and talked with polar explorers and scientists from many countries. To her delight, many of them accepted her as an equal and shared their knowledge with her. Boyd made her own valuable contribution. She recorded the search on more than 20,000 feet (6,100 meters) of movie film and several thousand photographs; later she turned this unique pictorial history over to the American Geographical Society (AGS). King Haakon VII of Norway awarded Boyd the Order of St. Olav. She was the first non-Norwegian woman to receive this honor.

Now that she had taken part in polar history, Boyd wanted to make her next voyage to the Arctic a scientific journey rather than a purely recreational one. Dr. Isaiah Bowman, director of the AGS, suggested that she could pursue her interest in photography and learn the new science of photogrammetry—a technique that used special cameras to take photographs from which highly accurate, detailed maps could be made. Photogrammetry allowed the mapping of glaciers, mountain peaks, and other terrain that was difficult or impossible for surveyors to reach on foot.

Boyd had chosen a challenging destination for her first serious expedition. She planned to go to the eastern coast of Greenland, a huge island off the coast of North America that belongs to Denmark. This coast is washed by the East Greenland Current, which carries freezing water and masses of polar ice down from the Arctic Ocean. As a result the coast is frozen solid except for a few months every summer. Even during the summer its waters are hazardous because of icebergs that have broken loose from the polar ice pack or fallen from the Greenland glaciers. The French polar explorer Jean Baptiste Charcot described the East Greenland coast as "sinister, implacable, and often murderous."

The frontispiece of Boyd's *The Coast of Northeast Greenland* features the *Veslekari,* the sturdy Norwegian ship that made so many voyages to the grim Greenland coast. The publication of this book was delayed until 1948 because the American military authorities did not want the information it contained to fall into German hands during World War II.

The echo-sounding machine measured the depth of the sea and allowed scientists to piece together a picture of the peaks and crevasses of the sea floor. Boyd was one of the first explorers to use the echo-sounder; in *The Coast of Northeast Greenland* she showed how it was installed.

Yet this forbidding, almost inaccessible place is one of the most dramatic, magnificent coastlines in the world. Like the coasts of Norway, Iceland, and southern Chile, it is broken by dozens of fiords—deep, narrow, winding canyons in the coastal mountains, carved over millions of years by glaciers as these rivers of ice ground their slow way from the ice cap in the center of the country to the sea. The fiords of East Greenland form one of the world's most elaborate and complex fiord systems. Their general outlines had already been explored and mapped, chiefly by Scandinavian explorers. But Boyd had learned that she could make a genuine scientific contribution by producing a detailed survey of small but remote areas that not yet been fully explored. Her range would be from the 70th

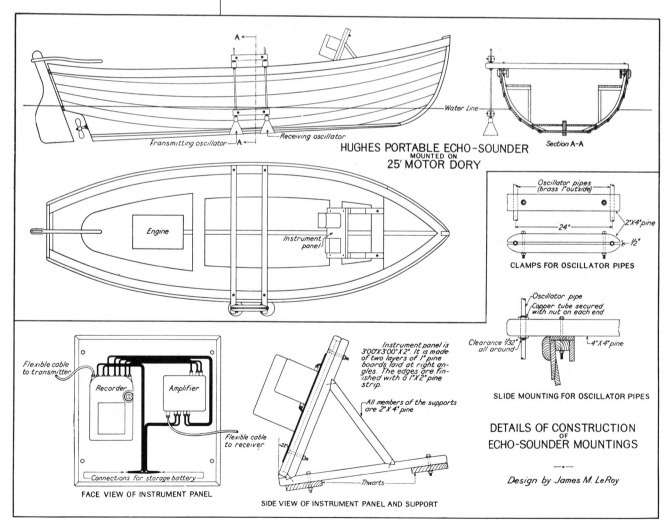

to the 74th parallels of latitude, especially Franz Joseph Fiord, King Oscar Fiord, and the many smaller fiords that branched from them—"the largest ramification of fiords in the world," as she said.

In other parts of the world, fiords are simple channels, although they may be quite long and crooked. But the fiords of East Greenland are often interconnected with cross-channels and bays, and parts of this maze are almost perpetually blocked with pack ice. East Greenland's fiords are also extremely deep, both above sea level and below it. The water in 120-mile-long (192-kilometer-long) Franz Joseph Fiord, the largest fiord in Boyd's area of study, is half a mile deep. Its steep rock walls are as tall as the Grand Canyon, although the fiord is much narrower.

Into this wilderness of rock and ice came Louise Boyd in the summer of 1931. She brought her camera equipment, her maid, and six friends aboard the *Veslekari,* a 125-foot (38-meter) sealing ship from Norway, and she enjoyed a brief but remarkably successful season of exploration.

While the *Veslekari* followed the coastline of every fiord in the region, Boyd took thousands of photographs of everything she saw, from the largest glacier to the smallest wildflower nestled between two pebbles. Her biggest triumph occurred in Ice Fiord, an arm of Franz Joseph Fiord. No one had ever gone all the way to the head of Ice Fiord, but Boyd succeeded in doing so, and she discovered there an enormous glacier that no one had reported before. She also made the first crossing of an unexplored valley between two fiords and found that they were connected by another large glacier, also previously unknown. The pictures she took to document these discoveries were so complete that the American Geographical Society was able to use them as the basis for new, accurate maps of the region. The Geographic Institute of Denmark named the area at the head of Ice Fiord "Miss Boyd Land" in her honor.

Before returning to Norway across the Greenland Sea and the Norway Sea, the *Veslekari* swung south and stopped at Scoresbysund, one of only two settlements on Greenland's east coast. Scoresbysund was a community of about ninety of Greenland's native Inuit people, originally from western Greenland. The village had been established by the Danish government just six years earlier in an attempt to populate the east coast. Its lonely inhabitants were delighted to see the *Veslekari,* and they greeted their visitors with a display of Inuit dancing and singing.

Boyd admired the "splendid, kindly" Inuit and observed their clothing, homes, boats, and way of life with interest. But her real passion was for the fiordland to the north, which haunted her with its beauty. The "imaginary gates" of the Arctic had swung open for her, and she was continually drawn back to explore the mysteries that lay beyond them.

Boyd was frequently asked why anyone—especially a woman—would want to return to the deadly, monotonously white Arctic. She would reply impatiently that the towering, brilliantly colored rock walls of the fiords, the endless plains covered with vivid wildflowers during the short northern summer, and the dark sea fringed with diamond-bright ice made "a picture of such majesty and on so vast a scale that no explanation need be given by any explorer for wishing to revisit such a scene."

And revisit it she did, in 1933, 1937, and 1938, each time in the *Veslekari.* The expedition of 1933 was a turning point, for it was sponsored by the American Geographical Society. The AGS agreed to provide Boyd with scientific and practical advice, as well as a staff of scientists. She paid all of the expedition's expenses, however, as she did on all of her later trips. As the expedition leader, Boyd would write its official report, which the AGS would publish. At last, Boyd felt, she had the opportunity to become a serious explorer.

The mission of the 1933 expedition was to study the rocks, plants, and landforms at the edges of glaciers in the fiord region, especially at the head of Franz Joseph Fiord. The team included two surveyors, a geologist, and a botanist. Boyd and several assistants (needed mostly to carry her heavy camera equipment) would expand the photogrammetric work she had begun on the previous trip. But the botanist developed appendicitis early in the voyage, and Boyd—who was an accomplished gardener and knew quite a lot about botany—took over his job of collecting plant specimens, doubling her own scientific contribution to the expedition.

She spared no expense in fitting out the *Veslekari.* She provided the latest and best equipment for all the scientists, including one of the newly invented echo-sounding machines that measured the depth of the sea. She had the echo-sounder installed in a special compartment built into the ship's hull. She also equipped the expedition with two cooks, a library, and lavish provisions—so lavish that the scientists celebrated July 4 by eating five pounds of costly caviar.

But although she enjoyed her little luxuries, Boyd was quite earnest about her scientific purpose. She ran the expedition with an iron hand and expected her orders to be followed to the letter. She herself worked very hard without complaint, and she felt that others should do the same. On one occasion she sent two of the scientists inland to examine a particular area. The goal she had assigned was a long way away, however, and the two men decided that no real harm would be done if they turned back early. They did so, but told Boyd that they had completed the trip. To their dismay, she decided to visit the same place a few days later. They waited apprehensively on the ship, knowing that she would see from their footprints in the snow that they had lied to her. When her determined little figure trudged into view in the distance, one of them said, "This is where we'd better duck." Her reaction was not recorded, but the two were not invited on the next expedition. Nothing made Boyd angrier than being treated like a fool.

The 1933 expedition was full of accomplishments and exciting incidents. Among other things, they discovered that the echo-sounder could be used to find schools of fish—a discovery that was promptly put to use by commercial fishing fleets around the world. Boyd and a member of the crew made a side trip to a previously unexplored area where they made an eerie find one evening while preparing to photograph the icebergs in the fiord. They saw a bottle bobbing in the cold dark water—a startling sight in a place so far from any human habitation. They fished it out and found that it held a note: "This point, the inmost of Franz Joseph Inlet, I have reached at the 12 August 1899, alone with a canoe: Dr. Josef Hammar, The Swedish Greenland Expedition 1899." The bottle had been floating in the fiord for more than thirty years, a hopeless message from a member of a Swedish expedition that had disappeared while trying to cross the North Pole by balloon.

Boyd's own expedition nearly ended in disaster. In early September, as winter began to move in, the expedition members made their final, hasty studies and prepared to leave the fiords for the open sea.

But the *Veslekari* ran aground on a mudbank. Suddenly Boyd and her companions were faced with the dreaded possibility of being frozen in for the long Arctic winter. In several days of frantic labor, they managed to lighten the ship by abandoning its reserves of coal, fuel oil, and water, as well as its three lifeboats. Then they fastened a cable from the ship to an iceberg that was floating out to sea on the tide, and hoped for the best. Luckily, the ship was now light enough for the berg to tow it off the mudbank. The crew quickly recovered the lifeboats and fuel oil and steamed off down the fiord.

The delay, however, had been costly. The first winter gales were already beating against the shores of East Greenland, and the *Veslekari* was driven back twice by snow and raging 55-mile-an-hour (88-kilometer-an-hour) winds. On the third try, they succeeded in getting away from the dangerous coast.

The results of the expedition were highly satisfactory to the American Geographical Society and to Boyd. Her report was published as *The Fiord Region of East Greenland* and illustrated with 350 of her photographs. From the information she and her colleagues had gathered, the AGS was able to make four new maps of the glaciers and valleys around the fiords, as well as maps of the sea floor in Franz Joseph Fiord and between Norway and Greenland.

During World War II the U.S. Coast Guard put Boyd's knowledge of the northern waters to good use. She is shown here on a mission to study radio communications in the Arctic.

The AGS sponsored Boyd's 1937 and 1938 expeditions to the fiord region. She and the scientists whom she invited to accompany her brought back new information each time, but her greatest achievement as an explorer came at the end of the 1938 season. Hoping to set a new record for the northernmost point reached by a ship on the East Greenland shore, she ordered the *Veslekari* to go as far north as possible before the ice closed in.

The ship reached the tiny island of Ile de France, where Boyd and several other members of the team went ashore for several hours to take photographs. But a thick pack of ice halted the *Veslekari* at 77°48' north latitude, only 800 miles (1,280 kilometers) from the pole. Boyd wrote, "There we lay drifting in the ice through the night and early morning of August 3, hoping that the ice would change so that we would be able to continue northward, but it was obvious next morning that we had gone as far north as we could go." Regretfully they turned south for home, but Boyd spent many hours on deck with her cameras in the bitter cold, "in order to get as full a record as possible of this seldom-visited coast."

She had indeed set a new record. The AGS reported to the newspapers that Boyd had "gone farther north in a ship along the East Greenland shore than any other American." A ship from a French expedition had reached a point slightly farther north in 1905, but its crew had been unable to land. So Boyd could claim to have made the northernmost landing along the coast.

Boyd's reports of the 1937 and 1938 expeditions were to have been published in 1940 as *The Coast of Northeast Greenland*. But with the outbreak of World War II in 1939, the United States government feared that Boyd's maps and information might be useful to the enemy, Germany, which was building bases in the Arctic. At the government's request she not only delayed publication of her report until 1948 but also turned over her records to the U.S. Army. She made a daring, last-minute dash to Norway, just as that country was falling to the Germans, to bring back the records and equipment that she had stored there.

During the war, Boyd served as an adviser to the government on Arctic conditions. She was part of a U.S. Coast Guard expedition to the Canadian Arctic to study the effects of polar magnetism on radio transmissions. The captain of this expedition was Robert Bartlett, a veteran of many trips to the polar ice pack and a longtime friend of Peary and other polar explorers. He made it clear that he resented the presence of a "society woman" on his ship, and Boyd found the whole experience humiliating.

Boyd had encountered prejudice like Bartlett's throughout her career. At the time she began her Arctic expeditions, women had begun to travel fairly extensively in many parts of the world, but polar exploration remained a small, elite, male club. Boyd was the first woman to demand admission. In fact, she bought her way in. If she had not been able to hire her own ships and equip her own expeditions, she would probably not have received any help from governments or research organizations. Eventually, however, most polar explorers came to regard her as a colleague and equal, although there were still some, like Bartlett, who felt that a woman had no place in the Arctic.

Some scientists, too, looked down on Boyd as nothing but a rich, high-handed amateur with no real scientific background—no college degree and no academic training. But although her money may have helped her attract the interest of the American Geographical Society at first, she proved herself to be much more than a thrill-seeker or a dabbler. She spent a great deal of time with experts in photography and botany, learning the proper techniques for filming and specimen collecting, and she delivered results that were not only respectable but had real scientific value.

By the time World War II was over, Boyd was nearly sixty. She devoted the next decade to working for charitable and civic groups in her native San Rafael. She made no further sea voyages to Greenland, but she had one goal still to achieve. She wanted to see the North Pole. In 1955, at the age of sixty-eight, she chartered a plane and flew across the pole from Norway. She was accompanied by Norwegian General Finn Lambrecht. The flight was something of a reunion for these two polar explorers; years before, during the search for Amundsen, he had been one of the officers stationed with her on the *Hobby*.

Boyd's flight was the first private, nonmilitary flight across the pole and also the first by a woman. In a telegram to the Society of Women Geographers, of which she had been a member for many years, Boyd said: "Flew over North Pole yesterday morning at nine fifteen and circled the Pole. Did so in brilliant sunshine and cloudless blue sky and perfect visibility....Greetings to all women geographers."

Boyd never returned to the Arctic, but she continued to spend nearly half of every year traveling. Her journeys were typically ambitious and exhausting. In 1958, for example, she went around the world, stopping in Japan, Hong Kong, Macao, Vietnam, Thailand, Cambodia, India, Pakistan, Afghanistan, and Turkey.

In 1962 Boyd moved from her beloved Maple Lawn estate to an apartment in San Francisco because she could no longer afford to maintain the estate. Her fortune had been drained by a lifetime of travel, exploration, fine living, and large donations to charity and the arts. She spent her last few years in a nursing home, where she died in 1972, at the age of eighty-five, of cancer.

In the years before her death, Louise Arner Boyd received the recognition she had craved and earned. In 1967 she was a guest of honor at a dinner of the prestigious Explorers Club in New York, where she was introduced as "one of the world's greatest woman explorers." Three colleges awarded her honorary degrees. She was the first woman to serve on the Council of the American Society of Geographers. She was elected a member of Britain's Royal Geographical Society. And the American Polar Society declared that she had contributed more to the knowledge of Spitsbergen, Franz Joseph Land, East Greenland, and the Greenland Sea "than any other explorer." Thanks to Louise Arner Boyd, the Arctic was no longer only a man's world.

Between voyages to the Arctic, Boyd lived the life of a wealthy socialite. She gave elegant parties at her California estate and dressed in fashionable clothes and furs. Sadly, she ran out of money and had to spend her final years in humble surroundings.

CHAPTER 9

Freya Stark: The Art of Travel

J ust before her fourth birthday, Freya Madeline Stark ran away from her home in Dartmouth, England. Equipped with nothing more than a raincoat, a toothbrush, and a penny, she set out for Plymouth, a port on the English coast. She was determined to go to sea. She did not get very far on that attempt, but Freya Stark never lost her determination to travel. Later in her life she became one of the most notable travelers of the twentieth century.

Stark was born in Paris in 1893. Her father was English; her mother was of mixed English and Italian descent and had grown up in Italy. Her parents traveled around Europe a great deal when Stark was small. She was carried in a basket over the Dolomite Mountains of Italy while she was still a baby, and some of her earliest memories are of landscapes rolling past outside train windows.

For the first eight years of Stark's life, home was the English countryside. In 1901, her mother and father separated. Stark and her younger sister Vera went to live in the small town of Asolo, Italy, with their mother. Her father remained in England for a time and then settled in British Columbia, Canada.

Stark's adolescence was troubled by the family's occasional money worries and also by frequent disagreements between her and her mother. Furthermore, her health was not good. Like Isabella Bird, she suffered from recurring illnesses all her life: pneumonia, fevers, heart problems. But like the dauntless Bird, she refused to allow illness

Englishwoman Freya Stark, shown wearing an Arab head-dress, became one of the twenti-eth century's most notable travelers. Her greatest achieve-ments as an explorer were her journeys in little-known corners of Persia, Arabia, and Turkey.

to limit her horizons. Whenever she was confined to bed or to home by a bout of poor health, she studied a new language and made plans for her next trip.

As a teenager Stark suffered a terrible injury that scarred her for life. Her long hair was caught in a machine in her mother's carpet factory, and before the machine could be stopped Stark's scalp was badly torn. Ever afterward she was self-conscious about the scars on one side of her face and about her looks in general. Although she sometimes wished she were prettier and taller, she learned to take great delight in stylish, colorful clothes—in later years large, unusual hats were her trademark.

Stark entered a happy, lively period in 1912, when she went to London to study history at the University of London. But before she could complete her degree, World War I broke out, and she returned to Italy to be near her mother. During the war she worked for the Allied government in Italy as a mail censor and a nurse. At this time she fell deeply in love with a young Italian man. They became

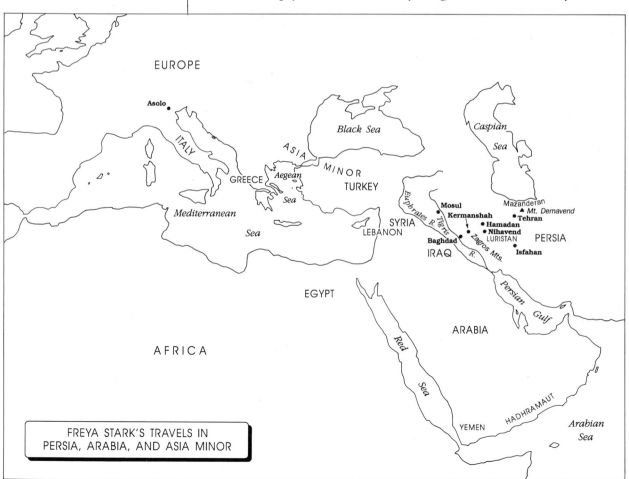

FREYA STARK'S TRAVELS IN
PERSIA, ARABIA, AND ASIA MINOR

engaged, but to her lasting sorrow he broke the engagement and married another woman.

After the war Stark was unhappy about her broken engagement and bored with life in Asolo. She tried her hand at fruit-farming and other moneymaking schemes, but she remained restless. Finally she decided that what she really wanted to do was travel. She longed to see the world. And because her only money was a small sum given to her by her father, she knew that she could not travel as a pampered tourist. She would have to rough it on her own, with no luxuries. Because she was particularly interested in the Muslim countries of the Near East and Arabia, she decided to learn Arabic to make travel easier. She found a monk who had lived in Lebanon and was willing to teach her, so twice a week she walked for an hour to the train station, rode to the town where the monk lived, and then walked a few miles to the monastery. She was a diligent student; within a few months she was reading the Koran, the holy book of Islam, in the original Arabic.

Her first venture into Muslim lands was a visit to Syria in 1927, when she was thirty-four years old. The following year she went to Lebanon, and this trip determined the direction of her life. She fell in love with desert travel, a passion that she was to pursue for years. As she wrote to a friend, "Yesterday was a wonderful day: for I discovered the Desert!...I never imagined that my first sight of the desert would come as such a shock of beauty and enslave me right away." Later in the same trip she wrote of the peaceful pleasures of camping in the desert, "with the camels tethered outside...and the sense of great spaces around us, and the silence and the nearness of the stars." This was a pleasure that never faded for Freya Stark. Thirty years later she traveled again through the desert around the Tigris River, and she wrote, "I was happy to be out in the wild and open world, with night and the long-unaccustomed slight spice of danger. Darkness fell at six-thirty, but the moon rose behind us, and trees and shrubs, distorted into strange sub-human shapes in the twilight, swam out clear into loveliness, as if their earth had crumbled to gold."

Her travels in Syria and Lebanon, with their combination of deserts, mountains, and ancient ruins, whetted Stark's appetite for more travel in the Middle East. In 1929 she moved to Baghdad, the capital of Iraq. A number of foreigners, mainly British people, were living in Iraq at the time, working for the government or for international oil companies. They lived in an elegant quarter of the city and did not mix socially with the Iraqis. Stark could not afford to live that way—and she would not have wanted to even if she could have afforded it. She rented a room in the house of an Iraqi shoemaker and happily practiced her Arabic on his family. The other British residents in Baghdad were at first appalled by this outlandish behavior, but gradually Stark won over those who appreciated her wit, charm, kindness, and learning. Many of the people she met in Baghdad became lifelong friends and correspondents.

During the next few years she made a number of journeys in Iraq and neighboring Persia. Much of the time she was on ground that many other travelers had trod before her, but still she considered herself an explorer. *All* travel to new places, she felt, was a form of personal exploration. In 1930-32, however, she undertook some serious geographic exploration in two regions in Persia.

One region was Luristan, located in the Zagros Mountains of western Persia, north of where Marguerite Baker Harrison had traveled with the Bakhtiari a few years earlier. Luristan was inhabited by a nomadic people called the Lurs, about whom little was known. Stark found a guide who agreed to take her into Luristan, and she spent several weeks traversing steep crags and hidden valleys on horseback, visiting and photographing the Lurs. The nomads wore medieval clothing and seemed utterly isolated from the outside world; Stark was one of only a handful of Europeans to encounter them before their traditional tribal society was forever changed by contact with modern culture.

A little later she had an exciting but inconclusive adventure. At a party in Baghdad she heard about Hasan, a young Lur man who claimed to own a map showing the way to a secret cave in the hills near the city of Nihavend. Hasan said the cave held treasure: gold coins and statues and hundreds of precious jewels. He wanted a European accomplice to help him get the treasure out of the country. Stark met Hasan and believed his story. She decided to go after the treasure and smuggle it away; this was illegal, but she claimed it was better for the treasure to end up in a museum in London, where historians could study it, than to be melted down or squandered by

Stark in Mazanderan, following rumors and legends to abandoned castles. She was a willing horsewoman but sometimes grew impatient with steeds that seemed timid or unmanageable. In *The Valleys of the Assassins* she wrote, "An unwilling horse and a dragging child and a woman who insists on explaining her motives are the three most wearying objects in creation."

the Iraqis. Hasan gave her a copy of the map and promised to meet her near the cave.

Unfortunately, the Baghdad government had also heard rumors of the treasure of Nihavend. Police followed Stark closely all the way to the cave region, and she was unable to shake them off long enough to complete her search. Hasan did not show up; she decided the police had frightened him off. So Stark abandoned the quest without seeing the cave for herself. But when she published the story of her treasure hunt in a book called *The Valleys of the Assassins,* she disguised the supposed location of the cave—just in case the map was correct. She did not want anyone else to succeed where she had failed.

The Valleys of the Assassins also describes Stark's travels in another part of Persia called Mazanderan, in the Elburz Mountains on the southern shore of the Caspian Sea. At the time, Mazanderan was covered with magnificent forests of rare trees—it was one of the last large remnants of the forest that once covered all of prehistoric Europe and much of western Asia. Stark traveled through this beautiful, seldom-visited region looking for the ruins of mountaintop castles. She had become fascinated by the legends about the Assassins, a medieval Islamic cult of warrior mystics who were said to live on rugged peaks in Mazanderan. The region between the Caspian Sea and Tehran was what Stark called "a lovely blank on the map." She decided to fill it in—and to find some of the Assassins' castles while she was at it.

When she set out for Mazanderan from the city of Hamadan she wrote, "This is a great moment, when you see, however distant, the goal of your wandering....It matters not how many ranges, rivers, or parching dusty ways may lie between you: it is yours now forever." All her life Stark had felt that the beginning of a journey is a joyous occasion, one of life's finest experiences.

Stark did succeed in locating and exploring several mountaintop ruins, although she was gravely ill with dysentery and malaria. She collapsed and would have died but for the kindness of a local woman who looked after her for more than a week. But the results of her trip were impressive. The British War Office used information she had gathered to improve its maps of Mazanderan, and the Royal Geographical Society (RGS) of Britain gave her an award for her work in both Luristan and Mazanderan. Britain's Royal Asiatic Society awarded her its Burton medal, named for Sir Richard Burton, an Englishman who had explored in Persia, Arabia, and Africa in the nineteenth century; Stark was the first woman to receive this award.

A wedding in the Hadhramaut region of Arabia, photographed by Stark in 1935. She was one of the first Europeans to visit the ancient cities hidden in the valleys of southern Arabia.

Stark now realized, as Ida Pfeiffer had, that she could make a living by writing, especially if she wrote about travel. Her first book, *Baghdad Sketches,* was published in 1933 while she was working for a newspaper in Baghdad. *The Valleys of the Assassins* appeared the following year.

In 1935, when she was forty-two, Stark made an excursion into the Arabian Peninsula. She was curious about the part of Arabia that is called the Hadhramaut—the long southern coast that runs from the Red Sea to the Persian Gulf. For many centuries the Arab merchants of this region had operated a profitable trade in spices and incense from desert plants, and they were traditionally suspicious and wary of strangers. In addition, their strict adherence to Islam meant that they did not look with favor upon a woman traveling alone. But Stark was respectful of their customs and patient when she encountered obstacles. Eventually she was able to visit some of the age-old walled cities tucked into the valleys of the Hadhramaut. She was the first European woman to do so. The trip was cut short, though, when she suffered a heart attack and had to be rescued by the Royal Air Force. Stark described her Arabian adventures in *The Southern Gates of Arabia* (1936).

Soon afterward, with backing from the RGS, she accompanied two other British women travelers on a trip to excavate archaeological sites on Arabia's southwest coast, in the present-day country of Yemen. Already her love of travel was firmly linked to her interest in history. Yet the archaeological expedition was disappointing to Stark. She bickered constantly with the other two women, who felt that she wasted too much time sitting and chatting with the local people. She, on the other hand, felt that the archaeologists were unsociable souls who ignored the courtesies of the present while they single-mindedly sifted through broken bits of the past. "I hate archaeology if it means that one's whole soul has to turn into statistics and eliminate human beings," she wrote to a friend.

The expedition to Yemen showed Stark what she did *not* want to do in her writing. She loved history, and she reveled in the past of every place that she visited. But she would not allow the glories of the past to blind her to the everyday details of life in the present. She developed a writing style that weaves together the panorama of history, the distinctive character of landscapes, and the day-to-day lives of ordinary people. Her special gift as a travel writer is to make a place, its history, and its people come alive in a narrative of her own adventures.

Stark's expeditions in Persia and Arabia established her not only as an explorer but also as a historian and a writer. When World War II broke out in 1939, she went to work for the British Ministry of Information in Arabia, Egypt, and Iraq, writing pro-British newspaper articles and trying to build support among the Arab nations for Britain and its allies. One of her colleagues in the Ministry of Information was a man named Stewart Perowne. She and Perowne were married in 1947, but the marriage ended in 1951. By then Stark was supporting herself as a writer. Her headquarters was a house in Asolo that had been left to her by a family friend.

The next phase of Stark's travels began in 1952, when she was fifty-nine years old. She turned her attention to the countries that border the Aegean Sea and the northeastern corner of the Mediterranean Sea. The ancient Greeks and Romans called this part of the world Asia Minor, and today it consists of Greece, Turkey, and Syria. Over a fourteen-year period, Stark traveled extensively in Asia Minor and wrote many books of travel and history essays; she also wrote her autobiography.

Turkey became one of Stark's favorite places. It has been a crossroads of peoples, cultures, and religions since the dawn of history. Greeks, Persians, Central Asian nomads, Romans, and other peoples have left their mark on it, and in different eras it has been the capital of both Christian and Islamic empires. Stark once remarked that in Turkey "a journey without history is like a portrait of an old

Stark addressing a meeting of an Arab society in Cairo, Egypt, in 1940. During World War II she worked for the British propaganda effort, trying to win allies for Britain in the Islamic countries of the Middle East.

face without its wrinkles"—a false or superficial portrait. She loved to venture off the beaten track in Turkey, perhaps looking for a village or a mountain that was mentioned by a Greek historian more than two thousand years earlier. One of her best travel books from this period is *Alexander's Path,* which traces the route that Alexander the Great once followed through Turkey.

But Stark's absorption with the past never blinded her to the present: All of her travel books present clear, perceptive pictures of the people among whom she traveled. No matter where she went, she seemed to have a sympathetic understanding of people's concerns and activities and ways of life.

Stark's fame both as a traveler and a writer continued to grow. In 1972, Queen Elizabeth II of England made her a Dame of the British Empire (which is the women's equivalent of being knighted). Stark once said, "Curiosity ought to increase as one gets older." Hers did. She went to China for the first time when she was in her seventies, she toured a remote part of Afghanistan by jeep at seventy-six, and a year or so later she trekked by pony in the Himalayan kingdom of Nepal.

All her life Stark has understood what she calls "the art of travel"— an art that can be practiced by anyone on any journey. Here is how she described it in *Baghdad Sketches,* her first book:

> To awaken quite alone in a strange town is one of the pleasantest sensations in the world. You are surrounded by adventure. You have no idea of what is in store for you, but you will, if you are wise and know the art of travel, let yourself go on the stream of the unknown and accept whatever comes in the spirit in which the gods may offer it. For this reason your customary thoughts, all except the rarest of your friends, even most of your luggage—everything, in fact, which belongs to your everyday life, is merely a hindrance. The tourist travels in his own atmosphere like a snail in his shell and stands, as it were, on his own perambulating doorstep to look at the continents of the world. But if you discard all this, and sally forth with a leisurely and blank mind, there is no knowing what may not happen to you.

AFTERWORD

The Journey Continues

The spirit of exploration is alive and thriving in women today. In 1963 Dervla Murphy, an Irishwoman with an itch to see the world, embarked on the first of a series of journeys that echo Isabella Bird and Fanny Bullock Workman. "I thought then, as I still do, that if someone enjoys cycling and wishes to go to India, the obvious thing is to cycle there," she wrote in *Full Tilt: Ireland to India with a Bicycle.* In *Eight Feet in the Andes* she described her travels in South America with a ten-year-old daughter and a pack mule. She has also adventured in Madagascar, Iran, and Ethiopia. In 1985 Sorrel Wilby, a twenty-four-year-old Australian woman, set out with $200 in her pocket to walk 1,800 miles (2,880 kilometers) from western Tibet to Lhasa—a pilgrimage of which Alexandra David-Neel would have approved. And the American writer Mary Morris described her travels in Mexico and Central America in her 1988 book *Nothing to Declare,* an account of the pleasures and perils of solitary travel.

The twentieth century has produced a host of women travelers, explorers, and adventurers. Some of them, like the aviator Amelia Earhart, the first woman to cross the Atlantic Ocean by air, were pioneers of new technology. Another technological pioneer is Sylvia Earle, an American scientist and diver who in 1979 reached the record-breaking depth of 1,250 feet (380 meters) beneath the sea's surface. Earle also led an all-woman team of scientists in an experiment in undersea living, spending two weeks in a submerged capsule in the Caribbean Sea. The success of the undersea project opened the way for women to be included in the American space program, so that Sally Ride and other American woman astronauts could join the company of explorers like Soviet cosmonaut Svetlana Savitskaya, the first woman to walk in space.

Science has taken twentieth-century women not just into the sea and into space but into the far places of the world. In 1924 Delia Denning Akeley went to Central Africa to collect wildlife specimens for the Brooklyn Museum of Arts and Sciences. She ventured into the Ituri rain forest in what is now Zaire and lived for several months with Pygmy tribespeople. More recently, two women naturalists have done much to enlarge the world's understanding of Africa and its wildlife: Dian Fossey with her study of the endangered mountain gorillas, and Jane Goodall with her study of chimpanzees.

Antarctica, too, now has its share of women meteorologists, biologists, and geologists. In 1947 a U.S. Navy officer named Harry Darlington told his wife, "There are some things women don't do. They don't become Pope or President—or go down to the Antarctic." But Darlington's wife, Jennie, spent a year in Antarctica; she was the first woman to do so. Soviet and Argentinian women scientists worked in Antarctica during the 1950s and 1960s, and in 1969 the U.S. Navy finally lifted its ban on women in the Antarctic Research Program. Today women are part of nearly all polar exploration and research teams.

Some twentieth-century women travelers have been record-setting sportswomen like England's Naomi James, who sailed around the world alone in nine months in 1977-78. Mountain-climbing also has continued to attract women since the days of Fanny Bullock Workman and Annie Smith Peck. In 1975 a Japanese team became the first all-women expedition to climb Mt. Everest, and a few years later an American women's expedition climbed Annapurna, another peak in the Himalayas.

In 1968, reflecting on her many decades of travel and adventure, Freya Stark wrote, "The lure of exploration still continues to be one of the strongest lodestars of the human spirit, and will be so while there is the rim of an unknown horizon in this world or the next." There will always be a frontier of some sort to explore—and women like Freya Stark and her sisters in adventure will always be ready to explore it.

Chronology

IDA PFEIFFER

1797
Born in Vienna, Austria

1842
Makes first trip abroad, to the Middle East; she follows it with a trip to Iceland

1846
Begins first journey around the world, in which she explores Brazil, China, India, Persia (Iran), and Russia

1851
Begins second journey around the world, in which she explores Borneo, Sumatra, Peru, and Ecuador

1857
Visits Madagascar and becomes involved in an attempt to overthrow Queen Ranavalona

1858
Dies in Vienna

ISABELLA BIRD BISHOP

1831
Born in Yorkshire, England

1854
Makes her first trip abroad, to the United States and Canada

1872
Visits the Sandwich Islands (Hawaii)

1878
Visits Japan, China, Southeast Asia, and Malaysia

1889-91
After the death of her husband, travels in India, Ladakh, and Persia

1894-97
Travels in Japan, Korea, and China

1901
Crosses Morocco

1904
Dies in Scotland

FLORENCE BAKER

1841
Born Florence Barbara Maria von Sass in Transylvania

1848
Family is killed and she becomes a refugee

1859
Samuel Baker buys Florence von Sass at a slave auction in Widdin, Bulgaria

1861-65
With Samuel Baker, explores the upper Nile River districts in Africa; they become the first Europeans to see Lake Albert

1870-73
Lady Florence and Sir Samuel Baker make a second expedition to the Lake Albert region

1893
Samuel Baker dies in England

1916
Florence Baker dies in England

FANNY BULLOCK WORKMAN

1859
Born in Worcester, Massachusetts

1881
Marries William Hunter Workman

1889
Moves to Europe

1895
Takes up bicyling

1897
Begins a series of bicycle journeys through India and South Asia that will total 14,000 miles (22,400 kilometers)

1898-1912
Makes eight mountaineering expeditions to the Himalaya and Karakoram mountains; sets her first world mountain-climbing record for women in 1899

1906
Sets a world climbing record for women of 22,815 feet (6,958 meters)

1925
Dies in France

MARY KINGSLEY

1862
Born in London, England

1892
Visits the Canary Islands and decides to travel in West Africa

1893-94
Makes her first journey to West Africa and collects river fish

1895
Explores the Ogowé River in West Africa and travels through the unknown country of the Fang people

1900
Dies while nursing prisoners of war in South Africa

ALEXANDRA DAVID-NEEL

1868
Born near Paris, France

1911
Goes to India; remains in Asia for 14 years, traveling and studying Buddhism

1915
Makes her first journey into Tibet

1924
Becomes the first European woman to reach Lhasa, the capital of Tibet

1936-44
Lives in China

1969
Dies in France

MARGUERITE BAKER HARRISON

1879
Born in Baltimore, Maryland

1918
Approaches the Military Information Division of the U.S. War Department for a job as a spy

1920-21
Imprisoned in the Soviet Union

1922
Travels across the Pacific Ocean and through Japan and Mongolia, then enters the Soviet Union and is imprisoned for a second time

1924-25
With Merian C. Cooper and Ernest Schoedsack, travels with the nomadic Bakhtiari people of Persia and makes an epic film of their annual migration

1925
With Blair Niles, Gertrude Emerson, and Gertrude Mathews Selby, forms the Society of Women Geographers

1967
Dies in Baltimore, Maryland

LOUISE ARNER BOYD

1887
Born in San Rafael, California

1924
Visits Spitsbergen, Norway, and determines to devote herself to Arctic exploration

1928
Helps search for missing polar explorer Roald Amundsen

1931-38
Organizes and leads four expeditions to East Greenland to make maps, take photographs, and collect plant specimens

1955
Becomes the first woman to fly over the North Pole

1972
Dies in San Francisco

FREYA STARK

1893
Born in Paris, France

1927-32
Makes a series of journeys in Lebanon, Iraq, and Iran

1935-38
Travels in southern Arabia

1952-65
Travels in Greece, Turkey, and the Near East

OTHER WOMEN TRAVELERS AND EXPLORERS

381-84

Etheria (or Egeria), a nun from Spain or France, travels to Jerusalem and Egypt; she writes a guide for pilgrims to the Holy Land

1805-06

Sacagawea, a Native American of the Shoshone people, helps guide the Lewis and Clark expedition across western North America to the Pacific Ocean

1832

Susanna Moodie emigrates from England to live in Ontario, Canada

1835

Emily Eden travels to India with her brother, Lord Auckland

1924

Delia Denning Akeley goes to Central Africa to collect wildlife specimens

1932

Amelia Earhart is the first woman to fly solo across the Atlantic Ocean

1963

Dervla Murphy sets off from Ireland to bicycle to India

1969

The U.S. Navy admits women to the Antarctic Research Program

1975

A Japanese women's climbing expedition reaches the summit of Mt. Everest

1979

Diver Sylvia Earle reaches a record depth of 1,250 feet (381 meters)

1984

Soviet cosmonaut Svetlana Savitskaya becomes the first woman to walk in space

1985

Sorrel Wilby treks from western Tibet to Lhasa

Further Reading

ABOUT WOMEN TRAVELERS AND EXPLORERS IN GENERAL

Aitken, Maria. *A Girdle Round the Earth: Adventuresses Abroad.* London: Constable, 1987.

Allen, Alexandra. *Travelling Ladies.* London: Jupiter, 1980.

Birket, Dea. *Spinsters Abroad: Victorian Lady Travellers.* Oxford, England: Blackwell, 1989.

De Pauw, Linda Grant. *Seafaring Women.* Boston: Houghton Mifflin, 1982.

Dole, Gertrude. *Vignettes of Some Early Members of the Society of Woman Geographers.* Closter, N.J.: Society of Woman Geographers, 1970.

Hamalian, Leo, editor. *Ladies on the Loose: Women Travellers of the 18th and 19th Centuries.* New York: Dodd, Mead, 1981.

Keay, Julia. *With Passport and Parasol: The Adventures of Seven Victorian Ladies.* London: BBC Press, 1989.

LaBastille, Anne. *Women and Wilderness.* San Francisco: Sierra Club, 1980.

Land, Barbara. *The New Explorers: Women in Antarctica.* New York: Dodd, Mead, 1981.

Lomax, Judy. *Women of the Air.* London: John Murray, 1986.

Melchett, Sonia. *Passionate Quests: Five Modern Women Travelers.* London: Faber & Faber, 1992.

Miller, Luree. *On Top of the World: Five Women Explorers in Tibet.* Seattle: Mountaineers, 1984.

Mills, Judy. "Great Explorations: Women Explorers." *Ms.,* June 1988, pages 58-63.

Olds, Elizabeth Fagg. *Women of the Four Winds: The Adventures of Four of America's First Women Explorers.* Boston: Houghton Mifflin, 1985.

Rappaport, Doreen. *Living Dangerously: American Women Who Risked Their Lives for Adventure.* New York: HarperCollins, 1991.

Rittenhouse, Mignon. *Seven Women Explorers.* Philadelphia: Lippincott, 1964.

Robertson, Janet. *The Magnificent Mountain Women: Adventures in the Colorado Rockies.* Lincoln: University of Nebraska Press, 1990.

Robinson, Jane. *Wayward Women: A Guide to Women Travellers.* New York: Oxford, 1990.

Russell, Mary. *The Blessings of a Good Thick Skirt: Women Travellers and Their World.* London: Collins, 1986.

Shapiro, Laura. "Fearless Spirits: Tales of Women Explorers." *Newsweek,* August 3, 1987, page 65.

Stevenson, Catherine. *Victorian Women Travel Writers in Africa.* Boston: Twayne, 1982.

Tinling, Marion. *Women into the Unknown: A Sourcebook on Women Explorers and Travelers.* Westport, Conn.: Greenwood Press, 1989.

BY AND ABOUT THE WOMEN WHO APPEAR IN THIS BOOK

Akeley, Delia. *Jungle Portraits.* New York: Macmillan, 1930.

Baker, Anne, editor. *Morning Star: Florence Baker's Diary of the Expedition to Put Down the Slave Trade on the Nile, 1870-73.* London: William Kimber, 1972.

Barr, Pat. *A Curious Life for a Lady: The Story of Isabella Bird, a Remarkable Victorian Traveler.* New York: Doubleday, 1970.

Bird, Isabella. *The Golden Chersonese and the Way Thither.* Kuala Lumpur: Oxford, 1967. Originally published in 1883.

———. *A Lady's Life in the Rocky Mountains.* Introduction by Daniel J. Boorstin. Norman, Okla.: University of Oklahoma Press, 1960. Originally published in 1879.

———. *Six Months in the Sandwich Islands.* Rutland, Vt.: Tuttle, 1990. Originally published in 1875.

———. *Unbeaten Tracks in Japan.* Rutland, Vt.: Tuttle, 1990. Originally published in 1880.

———. *The Yangtze Valley and Beyond: An Account of Journeys in China.* Introduction by Pat Barr. Boston: Beacon, 1987. Originally published in 1899.

Birkett, Deborah. "West Africa's Mary Kingsley." *History Today,* May 1987, pages 10-16.

Blum, Arlene. *Annapurna: A Woman's Place.* San Francisco: Sierra Club Books, 1980.

Boyd, Louise Arner. *The Coast of Northeast Greenland.* New York: American Geographical Society, 1948.

———. *The Fiord Region of East Greenland.* New York: American Geographical Society, 1935.

Brown, Marion. *Sacagawea: Indian Interpreter to Lewis and Clark.* Chicago: Childrens Press, 1988.

Cavan, Seamus. *Lewis and Clark and the Route to the Pacific.* New York: Chelsea House, 1991.

Conley, Andrea. *Window on the Deep: The Adventures of Underwater Explorer Sylvia Earle.* New York: Watts, 1992.

David-Neel, Alexandra. *My Journey to Lhasa.* Boston: Beacon, 1986. Originally published in 1927.

Dorrance, John. "Lady of the Arctic: Louise Arner Boyd." *San Francisco,* December 1981, pages 11-18.

Earhart, Amelia. *The Fun of It: Random Records of My Own Flying and of Women in Aviation.* New York: Putnam, 1932.

Earle, Sylvia. *Exploring the Deep Frontier: The Adventure of Man in the Sea.* Washington: National Geographic Society, 1980.

Eden, Emily. *Up the Country: Letters from India.* London: Virago Press, 1983. Originally published in 1866.

Fossey, Dian. *Gorillas in the Mist.* Boston: Houghton Mifflin, 1983.

Foster, Barbara, and Michael Foster. *Forbidden Journey: The Life of Alexandra David-Neel.* San Francisco: Harper & Row, 1987.

Frank, Katherine. *A Voyager Out: The Life of Mary Kingsley.* New York: Ballantine, 1987.

———. "Mary Kingsley: Victorian Adventurer in Africa." *Ms.,* February 1985, pages 22-27.

Goodall, Jane. *In the Shadow of Man.* London: Collins, 1971.

———. *Jane Goodall's Animal World.* New York: Atheneum, 1989.

————. *Through a Window: My Thirty Years with the Chimpanzees of Gombe*. Boston: Houghton Mifflin, 1990.

Harrison, Marguerite Baker. *Marooned in Moscow: The Story of an American Woman Imprisoned in Russia*. New York: Doran, 1921.

————. *There's Always Tomorrow: The Story of a Checkered Life*. New York: Farrar & Rinehart, 1935. Published in Great Britain as *Born for Trouble* (1936).

Howard, Harold P. *Sacajawea*. Norman: University of Oklahoma Press, 1972.

Kingsley, Mary. *Travels in West Africa*. Edited by Elspeth Huxley. London: J.M. Dent, 1987. Originally published in 1897.

————. *West African Studies*. London: Cass, 1964. Originally published in 1899.

Lauber, Patricia. *Lost Star: The Story of Amelia Earhart*. New York: Scholastic, 1988.

Lovell, Mary. *The Sound of Wings: The Life of Amelia Earhart*. New York: St. Martin's, 1989.

Moodie, Susanna. *Roughing It in the Bush; Or, Life in Canada*. London: Richard Bentley, 1852.

Moorehead, Caroline. *Freya Stark*. New York: Penguin, 1985.

Morris, Mary, *Nothing to Declare: Memoirs of a Woman Traveling Alone*. Boston: Houghton Mifflin, 1988.

Murphy, Dervla. *Eight Feet in the Andes*. Woodstock, N.Y.: Overlook Press, 1986.

————. *Full Tilt: Ireland to India with a Bicycle*. New York: Dutton, 1965.

————. *Muddling through in Madagascar*. Woodstock, N.Y.: Overlook Press, 1989.

O'Connor, Karen. *Sally Ride and the New Astronauts: Scientists in Space*. New York: Franklin Watts, 1983.

Orenstein, Peggy. "Champion of the Deep: Sylvia Earle." *New York Times Magazine,* June 23, 1991, pages 14-31.

Pfeiffer, Ida. *A Lady's Second Journey Round the World*. London: Longman, Brown, 1855.

————. *A Lady's Voyage Round the World*. London: Longman, Brown, 1851.

Ruthven, Malise. *Traveller through Time: A Photographic Journey with Freya Stark*. London: Viking, 1986.

Stark, Freya. *Alexander's Path*. Woodstock, N.Y.: Overlook Press, 1988.

———. *The Freya Stark Story*. New York: Coward-McCann, 1953.

———. *The Journey's Echo: Selected Travel Writings*. New York: Ecco, 1988.

———. *The Southern Gates of Arabia*. Los Angeles: Tarcher, 1983. Originally published in 1936.

———. *The Valleys of the Assassins*. Los Angeles: Tarcher, 1983. Originally published in 1934.

Wilby, Sorrel. "Nomads' Land: A Journey through Tibet." *National Geographic,* December 1987, pages 764-85.

Workman, Fanny Bullock, and William Hunter Workman. *Algerian Memories: A Bicycle Tour over the Atlas to the Sahara*. London: Unwin, 1895.

———. *In the Ice World of the Himalaya*. London: Unwin, 1900.

———. *Through Town and Jungle: Fourteen Thousand Miles A-Wheel Among the Temples and People of the Indian Plain*. London: Unwin, 1904.

———. *Two Summers in the Ice Wilds of the Eastern Karakoram*. London: Unwin, 1917.

Index

Rebecca Stefoff has written more than fifty books for young adults, specializing in geography and biography. Her longtime interest in reading and collecting travel narratives is reflected in such titles as *Lewis and Clark, Magellan and the Discovery of the World Ocean, Marco Polo and the Medieval Travelers, Vasco da Gama and the Portuguese Explorers, The Viking Explorers,* and numerous books on China, Japan, Mongolia, the Middle East, and Latin America. Ms. Stefoff has served as editorial director of two Chelsea House series, *Places and Peoples of the World* and *Let's Discover Canada*, and as a geography consultant for the *Silver Burdett Countries* series. She earned her Ph.D. at the University of Pennsylvania and lives in Philadelphia.